The Archaeological Heritage of Oman

MAGAN - THE LAND OF COPPER
Prehistoric Metallurgy of Oman

CLAUDIO GIARDINO

Sultanate of Oman
Ministry of Heritage and Culture

ARCHAEOPRESS PUBLISHING LTD
Summertown Pavilion
18-24 Middle Way
Summertown
Oxford OX2 7LG
www.archaeopress.com

© Claudio Giardino 2019

Magan - The Land of Copper, Prehistoric Metallurgy of Oman
(Includes bibliographical references and index).

1. Arabia. 2. Oman 3. Copper. 4. Prehistory 5. Antiquities.

Cover image: Ceremonial copper axe decorated with a bearded head and a lion found at 'Uqdat Al-Bakrah (photograph by Roman Garba, 2013)

First published in 2017 by the Ministry of Heritage and Culture, Sultanate of Oman, Muscat.
This edition is published by Archaeopress Publishing Ltd in association with the Ministry of Heritage and Culture, Sultanate of Oman.

Printed in England

ISBN 978-1-78969-178-8
ISBN 978-1-78969-179-5 (e-Pdf)

This publication is in copyright. Subject to statutory exception and to the provisions of relevant collective agreements, no reproduction of any part may take place without the written permission of the Ministry of Heritage and Culture, Sultanate of Oman.

Ministry of Heritage and Culture
Sultanate of Oman, Muscat

P.O. Box 668 P.C. 100
Khuwair, Muscat
Phone: +968 24 64 13 00
Fax: +968 24 64 13 31
Email: info@mhc.gov.om
Web Site: www.mhc.gov.om

Contents

List of illustrations and tables — vii

Acknowledgments — xv

Foreword — xvii

1 **A country of environmental diversities** — 1

2 **A geological and mineralogical overview** — 7

3 **Basic elements of metallurgy** — 12

4 **Copper for Sumer** — 20

5 **The earliest appearance of metalworking** — 29

6 **Early Bronze Age: the Hafit period, ca. 3200-2800 BC** — 40

7 **Early Bronze Age: the Umm an-Nar period, ca. 2800-2000 BC** — 63

8 **The prehistoric copper mines** — 84

9 **Copper smelting in prehistoric Oman** — 93

10 **Middle and Late Bronze Age: the Wadi Suq period. ca. 2000-1300 BC** — 106

11 **Early Iron Age, ca. 1300-300 BC** — 114

12 **Chemical-Physical Analyses by Energy Dispersive X-Ray Fluorescence (EDXRF)** — 139
Claudio Giardino and Giovanni Paternoster

Bibliographical references — 165

Index — 175

List of illustrations and tables

FIGURES

1.1.	Date palms at the oasis of Bat (photograph by C. Giardino).	2
1.2.	Cliffs at Ras Al-Jinz (photograph by C. Giardino).	3
1.3.	The coast at Ras Al-Hadd (photograph by C. Giardino).	4
1.4.	Mountains near Al-Moyassar (Maysar) (photograph by C. Giardino).	4
1.5.	Landscape of Rub Al-Khali, near 'Uqdat Al-Bakrah (photograph by C. Giardino).	5
1.6.	*Acacia* tree in the outer margins of the Rub Al-Khali (photograph by C. Giardino).	6
2.1.	Simplified geological map of Oman.	8
2.2.	Copper outcrop inside the Semail Ophiolite at Bilad Al-Maidin (photograph by C. Giardino).	10
3.1.	Photomicrograph of a copper-based object after casting showing a dendritic structure (after etching) (photograph by C. Giardino).	15
3.2.	Photomicrograph of a copper object after casting showing an equiaxial hexagonal grain structure (after etching) (photograph by C. Giardino).	15
3.3.	Photomicrograph of a copper object after hammering and annealing showing twinned grains (after etching) (photograph by C. Giardino).	16
3.4.	Microstructural changes in a metal submitted to mechanical and thermal treatments (after Giardino 2010).	16
3.5.	SEM image with elementary micro-mapping in a micro-structural feature.	18
3.6.	X-Ray Fluorescence portable equipment.	19
4.1.	Trade routes during the 3rd millennium BC, according to the seasonal winds.	21
4.2.	Reconstruction of an Early Bronze Age Magan boat (photograph by Helen Kirkbride, courtesy Oman Maritime Traditional Boat Centre).	24
4.3.	Diorite statue of Gudea of Lagash (courtesy Musée du Louvre).	26
5.1.	Metal objects from Wadi Shab GAS-1 (drawing by L. Tricarico).	32
5.2.	Metal objects from Wadi Shab GAS-1 (drawing by L. Tricarico).	32
5.3.	Ras Al-Hamra RH-10. Multi-tasking tool (DA 2672) (photographs by P. Koch, courtesy Oman National Museum).	34
5.4.	Ras Al-Hamra RH-10. Multi-tasking tool (DA 2672): arrows indicate wear and sharpening traces (drawing by L. Tricarico).	35
5.5.	Ras Al-Hamra RH-10. Detail of the tang of multi-tasking tool (DA 2672).	35
5.6.	Metal objects from Ras Al-Hamra RH-10 (drawing by L. Tricarico).	37
6.1.	Hafit cairn burials from Jebel Misht, Wadi Ayn (photograph by C. Giardino).	40

6.2.	Ras Al-Hadd HD-6. Three-dimensional reconstruction (image by M. Cattani, courtesy Joint Hadd Project).	42
6.3.	Ras Al-Hadd HD-6. Building 1 under excavation (photograph by M. Cattani, courtesy Joint Hadd Project).	42
6.4.	Ras Al-Hadd HD-6. Fishhooks (drawings by L. Tricarico).	46
6.5.	Ras Al-Hadd HD-6. Awls (drawings by L. Tricarico).	46
6.6.	Ras Al-Hadd HD-6. Chisels/punches (drawings by L. Tricarico).	47
6.7.	Ras Al-Hadd HD-6. Drill points (drawings by L. Tricarico).	48
6.8.	Ras Al-Hadd HD-6. Crochets/netting needles (drawings by L. Tricarico).	48
6.9.	Ras Al-Hadd HD-6. Shell-opener (drawings by L. Tricarico).	49
6.10.	Ras Al-Hadd HD-6. Pins (drawings by L. Tricarico).	51
6.11.	Ras Al-Hadd HD-6. Daggers (drawings by L. Tricarico).	52
6.12.	Ras Al-Hadd HD-6. Daggers (photograph by C. Giardino).	53
6.13.	Ras Al-Hadd HD-6. Blocklets/"flat pieces" (drawings by L. Tricarico).	54
6.14.	Ras Al-Hadd HD-6. Fragments for recycling (drawings by L. Tricarico).	55
6.15.	Ras Al-Hadd HD-6. Blocklet DA 16467. Distorted dendrites produced by cold hammering (SEM image).	56
6.16.	Ras Al-Hadd HD-6. Semi-finished pin DA 11585. Distorted dendrites produced by cold hammering (micrograph 5x).	56
6.17.	Ras Al-Hadd HD-6. Awl DA 15684.02. Twinned grains after hammering and annealing (micrograph 200x).	57
6.18.	Ras Al-Hadd HD-6. Whetstone DA 16471 (photograph by C. Giardino).	58
6.19.	Ras Al-Hadd HD-6. Stone-hammer DA 23106 (photograph by C. Giardino).	58
6.20.	Ras Al-Hadd HD-6. Reconstruction of the operative chain (drawings by L. Tricarico).	59
6.21.	Ras Al-Hadd HD-10. Long dagger with straight blade DA 14334 (photograph by P. Koch, courtesy Ministry of Heritage and Culture, Sultanate of Oman).	60
6.22.	Ras Al-Hadd HD-10. Awl DA 10979 (drawings by L. Tricarico).	61
6.23.	Ras Al-Jinz RJ-6. Long pin with small flatted head DA 8621 (photograph by C. Giardino).	61
7.1.	Umm an-Nar cairn burial from Bat (photograph by C. Giardino).	64
7.2.	Ras Al-Jinz, the area of RJ-2 (photograph by C. Giardino).	64
7.3.	Ras Al-Jinz RJ-2. Iron fishhook (DA 10958) (photograph by C. Giardino).	65
7.4.	Ras Al-Hadd HD-5 (surface). Copper seal (DA 12803) (photograph by P. Koch, courtesy Ministry of Heritage and Culture, Sultanate of Oman).	65
7.5.	Ras Al-Jinz RJ-2. Harappan style fishhook (DA 12792).	66
7.6.	Ras Al-Jinz RJ-2. Prehistoric structures and their chronology (image by V. Azzarà, courtesy Ras Al-Jinz Study Program).	67
7.7.	Ras Al-Jinz RJ-2. Flat ingot (DA 8506). Arrows mark the engraved line on its surface.	68
7.8.	Ras Al-Jinz RJ-2. Flat ingot Building I, Room 4 (DA 8506) and flat copper bar from Building I, Room 2 (DA 8366).	68

7.9.	Ras Al-Jinz RJ-2. Crucible fragments (RJ-2: 551,30-551,31). One crucible has its inner surface vitrified and green colored by copper (photograph by C. Giardino).	69
7.10.	Ras Al-Jinz RJ-2. Crucible fragments (RJ-2: 551,32-551,37). The inner surface of the crucible is badly damaged and vitrified by the high temperature of molten metal (photograph by C. Giardino).	69
7.11.	Ras Al-Jinz RJ-2. Small casting residue (DA 12673) (photograph by C. Giardino).	70
7.12.	Ras Al-Jinz RJ-2. ED-XRF spectrum from a crucible fragment (from QDF, U.S. 6062) containing a tin-copper alloy.	70
7.13.	Different fishhooks from Ras Al-Jinz RJ-2.	71
7.14.	Ras Al-Jinz RJ-2. Fishhooks (photograph by C. Giardino).	71
7.15.	Ras Al-Jinz RJ-2. Awls (photograph by C. Giardino).	72
7.16.	Ras Al-Jinz RJ-2. Chisel (photograph by C. Giardino).	73
7.17.	Ras Al-Jinz RJ-2. Crochets/netting needles (photograph by C. Giardino).	73
7.18.	Ras Al-Jinz RJ-2. Shell-openers (photograph by C. Giardino).	74
7.19.	Ras Al-Jinz RJ-2. Knives (photograph by C. Giardino).	75
7.20.	Ras Al-Jinz RJ-2. Flat axes (photograph by C. Giardino).	75
7.21.	Ras Al-Jinz RJ-2. Pins (photograph by C. Giardino).	76
7.22.	Ras Al-Jinz RJ-2. Rings/earrings (photograph by C. Giardino).	77
7.23.	Ras Al-Jinz RJ-2. Razors (photograph by C. Giardino).	78
7.24.	Ras Al-Jinz RJ-2. Necklace from Building VII with indication of the sampling spots for SEM/EDS analysis	78
7.25.	Ras Al-Jinz RJ-2. Blocklets/flat pieces (photograph by C. Giardino).	79
7.26.	Ras Al-Jinz RJ-2. Flat ingots (photograph by C. Giardino).	80
7.27.	Ras Al-Jinz RJ-2. Sheet or blades: fragment for recycling (photograph by C. Giardino).	80
7.28.	Ras Al-Hadd HD-7. Metal finds from tombs 4 and 5 (drawings by L. Tricarico).	81
7.29.	Ras Al-Jinz RJ-2. SEM/EDS spectra of the necklace from Building VII.	82
7.30.	Ras Al-Hadd HD-5. Cockle shell containing a metallic cosmetic amalgam (photograph by C. Giardino).	83
8.1.	Open cast mine at Al-Moyassar (Maysar) filled by mining debris (photograph by C. Giardino).	85
8.2.	Copper workings at Al-Moyassar (Maysar). Collapsed entrance of a mine (photograph by C. Giardino).	86
8.3.	Grooved stone-hammer from the mining area of Al-Moyassar (Maysar) (photograph by C. Giardino).	87
8.4.	Experimental firesetting (photograph by C. Giardino).	88
8.5.	Firesetting in Europe during 16th century (after Georgius Agricola, De Re Metallica Libri XII, Basel 1556).	89
8.6.	Bilad Al-Maidin. Rock cuts and broken ore fragments at a copper ore outcrop (photograph by C. Giardino).	90

8.7.	Linen clothes used as ventilation system in Europe during 16th century (after Georgius Agricola, De Re Metallica Libri XII, Basel 1556).	91
8.8.	Bilad Al-Maidin. Stone mortars and slag fragments scattered on the surface of the site (photograph by C. Giardino).	92
9.1.	Experimental smelting furnace (photograph by C. Giardino).	94
9.2.	Experimental smelting furnace loaded with charcoal (photograph by C. Giardino).	95
9.3.	Tapped slag fragment from Al-Moyassar-1 (DA 2219), with a smooth upper surface characterized by flow patterns (photograph by C. Giardino).	96
9.4.	Bilad Al-Maidin. Tapped slag on the surface of a slag heap (photograph by C. Giardino).	97
9.5.	Bilad Al-Maidin. Stone mortars for crushing slag and ore (photograph by C. Giardino).	97
9.6.	Maysar-1. Landscape (photograph by C. Giardino).	98
9.7.	Photomicrograph of a copper prill (2 mm diameter) imbedded in a slag from Al-Moyassar-1 (photograph by C. Giardino).	99
9.8.	Al-Aqir. Plano-convex ingot (DA 7406) (photograph by C. Giardino).	101
9.9.	Al-Aqir. Section of plano-convex ingot filled with slag (DA 7407) (photograph by C. Giardino).	102
9.10.	Copper ingot from Ras Al-Hadd HD-1 (upper and lower sides) (photograph by C. Giardino).	103
9.11.	Maysar-1. Photomicrograph of copper slag (DA 2219) containing copper prills (20 x) (photograph by C. Giardino).	104
10.1.	Cyprus, Mitsero (Nicosia). Ancient copper mines (photograph by C. Giardino).	107
10.2.	Cyprus, Mitsero (Nicosia). Ancient tapped copper smelting slag (photograph by C. Giardino).	107
10.3.	Plan of Ras Al-Jinz RJ-1 (after Cleuziou and Tosi 2007).	109
10.4.	Ras Al-Jinz RJ-1. Wadi Suq house during excavation (after Cleuziou and Tosi 2007).	110
10.5.	Metal objects from Ras Al-Jinz RJ-1.	111
10.6.	Casting residue from Ras Al-Jinz RJ-1. The arrow shows the cleaned point for the XRF analysis.	112
10.7.	Wadi Suq metal objects from the Al-Wasit grave (after Weisgerber 2007b).	112
11.1.	Metal objects from the tomb of Nizwa (after Al-Shanfari and Weisgerber 1989).	115
11.2.	The site of 'Uqdat Al-Bakrah in the sands of Rub Al-Khali (photograph by P. Koch).	117
11.3.	Dagger folded up for recycling (# 207): A, frontal view; B, lateral view (photographs by C. Giardino, modified by P. Koch).	118
11.4.	'Uqdat Al-Bakrah. Charcoal pit.	119
11.5.	'Uqdat Al-Bakrah. Furnace with stones around the perimeter.	119
11.6.	'Uqdat Al-Bakrah. Large casting residue that preserved as a cast the inside of a crucible (DA 27317 B (photograph by C. Giardino, modified by P. Koch).	120
11.7.	'Uqdat Al-Bakrah. Small casting residue (DA 26653.a) (photograph by C. Giardino, modified by P. Koch).	120

11.8.	'Uqdat Al-Bakrah. Semi-finished cast axes (photographs by C. Giardino, modified by P. Koch).	121
11.9.	'Uqdat Al-Bakrah. Dimpled stone tools emerging from the sand.	123
11.10.	'Uqdat Al-Bakrah. Stone anvil from the surface of the site.	123
11.11.	'Uqdat Al-Bakrah. Pierced whetstone (#170); on the left corner, detail of wear traces.	123
11.12.	'Uqdat Al-Bakrah. Fragment of a plano-convex copper ingot (DA 29943) (photograph by C. Giardino, modified by P. Koch).	124
11.13.	'Uqdat Al-Bakrah. Iron blocklets (DA 26655.b) (photographs by C. Giardino, modified by P. Koch).	124
11.14.	'Uqdat Al-Bakrah. Typology of the axes (photographs by C. Giardino, modified by P. Koch).	126
11.15.	'Uqdat Al-Bakrah. Typology of the hammers (photographs by C. Giardino, modified by P. Koch).	128
11.16.	'Uqdat Al-Bakrah. Typology of stakes (photographs by C. Giardino, modified by P. Koch).	130
11.17.	'Uqdat Al-Bakrah. Typology of hoes (photographs by C. Giardino, modified by P. Koch).	130
11.18.	'Uqdat Al-Bakrah. Typology of razors (photographs by C. Giardino, modified by P. Koch).	132
11.19.	'Uqdat Al-Bakrah. Typology of spatulas (photographs by C. Giardino, modified by P. Koch).	133
11.20.	'Uqdat Al-Bakrah. Typology of chisels (photographs by C. Giardino, modified by P. Koch).	134
11.21.	'Uqdat Al-Bakrah. Pair of shallow bowls used as scale plates (DA 29869 A, B) (photograph by C. Giardino) (photographs by C. Giardino, modified by P. Koch).	134
11.22.	'Uqdat Al-Bakrah. Typology of vessels (photographs by C. Giardino, modified by P. Koch).	135
11.23.	'Uqdat Al-Bakrah. Silver finger ring (DA 30197) (photograph by C. Giardino, modified by P. Koch).	136
11.24.	'Uqdat Al-Bakrah. Carnelian bead decorated with gold granulation (DA 26169.m). Top-left corner: micrograph of the granulation (photographs by C. Giardino, modified by P. Koch).	136
11.25.	Metal objects from the hoard of Ibri/Selme (after Weisgerber and Yule 1989; Yule and Weisgerber 2001).	137
12.1.	Schematic plan of the ED-XRF portable apparatus.	140
12.2.	Histogram of presence frequency of analyzed elements in objects from Wadi Shab GAS-1, Ras Al-Hamra RH-10, Ras Al-Hadd HD-6, Ras Al-Jinz RJ-2 and Ras Al-Jinz RJ-1.	153
12.3.	Copper concentrations in objects from Wadi Shab GAS-1, Ras Al-Hamra RH-10, Ras Al-Hadd HD-6, Ras Al-Jinz RJ-6, Ras Al-Hadd HD-10, Ras Al-Jinz RJ-2 and Ras Al-Jinz RJ-1, analyzed by XRF.	153
12.4.	Silver concentrations in objects from Wadi Shab GAS-1, Ras Al-Hamra RH-10, Ras Al-Hadd HD-6, Ras Al-Hadd HD-10, Ras Al-Jinz RJ-2 and Ras Al-Jinz RJ-6, analyzed by XRF.	154

12.5.	Zinc concentrations in objects from Wadi Shab GAS-1, Ras Al-Hamra RH-10, Ras Al-Hadd HD-6, Ras Al-Hadd HD-10, Ras Al-Jinz RJ-6, Ras Al-Jinz RJ-2 and Ras Al-Jinz RJ-1, analyzed by XRF.	154
12.6.	Iron concentrations in objects from Wadi Shab GAS-1, Ras Al-Hamra RH-10, Ras Al-Hadd HD-6, Ras Al-Hadd HD-10, Ras Al-Jinz RJ-6, Ras Al-Jinz RJ-2 and Ras Al-Jinz RJ-1, analyzed by XRF.	155
12.7.	Cobalt concentrations in objects from Wadi Shab GAS-1, Ras Al-Hamra RH-10, Ras Al-Hadd HD-6, Ras Al-Hadd HD-10, Ras Al-Jinz RJ-6, Ras Al-Jinz RJ-2 and Ras Al-Jinz RJ-1, analyzed by XRF.	155
12.8.	Arsenic concentrations in objects from Wadi Shab GAS-1, Ras Al-Hamra RH-10, Ras Al-Hadd HD-6, Ras Al-Hadd HD-10, Ras Al-Jinz RJ-6, Ras Al-Jinz RJ-2 and Ras Al-Jinz RJ-1, analyzed by XRF.	156
12.9.	Arsenic concentrations in tools from Wadi Shab GAS-1, Ras Al-Hamra RH-10, Ras Al-Hadd HD-6, Ras Al-Hadd HD-10, Ras Al-Jinz RJ-6, Ras Al-Jinz RJ-2 and Ras Al-Jinz RJ-1, analyzed by XRF.	156
12.10.	Arsenic concentrations in ornaments and weapons from Wadi Shab GAS-1, Ras Al-Hamra RH-10, Ras Al-Hadd HD-6, Ras Al-Hadd HD-10, Ras Al-Jinz RJ-6 and Ras Al-Jinz RJ-2, analyzed by XRF.	157
12.11.	Arsenic concentrations in semi-finished items from Wadi Shab GAS-1, Ras Al-Hamra RH-10, Ras Al-Hadd HD-6, Ras Al-Hadd HD-10, Ras Al-Jinz RJ-6, Ras Al-Jinz RJ-2 and Ras Al-Jinz RJ-1, analyzed by XRF.	157
12.12.	Nickel concentrations in objects from Wadi Shab GAS-1, Ras Al-Hamra RH-10, Ras Al-Hadd HD-6, Ras Al-Hadd HD-10, Ras Al-Jinz RJ-6, Ras Al-Jinz RJ-2 and Ras Al-Jinz RJ-1, analyzed by XRF.	158
12.13.	Nickel concentrations in tools from Wadi Shab GAS-1, Ras Al-Hamra RH-10, Ras Al-Hadd HD-6, Ras Al-Hadd HD-10, Ras Al-Jinz RJ-6, Ras Al-Jinz RJ-2 and Ras Al-Jinz RJ-1, analyzed by XRF.	158
12.14.	Nickel concentrations in ornaments and weapons from Wadi Shab GAS-1, Ras Al-Hamra RH-10, Ras Al-Hadd HD-6, Ras Al-Hadd HD-10, Ras Al-Jinz RJ-6 and Ras Al-Jinz RJ-2 and Ras Al-Jinz RJ-1, analyzed by XRF.	159
12.15.	Nickel concentrations in semi-finished items from Wadi Shab GAS-1, Ras Al-Hamra RH-10, Ras Al-Hadd HD-6, Ras Al-Hadd HD-10, Ras Al-Jinz RJ-6, Ras Al-Jinz RJ-2 and Ras Al-Jinz RJ-1, analyzed by XRF.	159
12.16.	Arsenic and nickel in objects from Wadi Shab GAS-1, Ras Al-Hamra RH-10, Ras Al-Hadd HD-6, Ras Al-Hadd HD-10, Ras Al-Jinz RJ-6, Ras Al-Jinz RJ-2 and Ras Al-Jinz RJ-1, analyzed by XRF.	160
12.17.	Arsenic and cobalt in objects from Wadi Shab GAS-1, Ras Al-Hamra RH-10, Ras Al-Hadd HD-6, Ras Al-Hadd HD-10, Ras Al-Jinz RJ-6, Ras Al-Jinz RJ-2 and Ras Al-Jinz RJ-1, analyzed by XRF.	160
12.18.	Lead concentrations in objects from Wadi Shab GAS-1, Ras Al-Hamra RH-10, Ras Al-Hadd HD-6, Ras Al-Hadd HD-10, Ras Al-Jinz RJ-6, Ras Al-Jinz RJ-2 and Ras Al-Jinz RJ-1, analyzed by XRF.	161
12.19.	Antimony concentrations in objects from Wadi Shab GAS-1, Ras Al-Hamra RH-10, Ras Al-Hadd HD-6, Ras Al-Hadd HD-10, Ras Al-Jinz RJ-2 and Ras Al-Jinz RJ-6, analyzed by XRF.	161

12.20.	Tin concentrations in objects from Wadi Shab GAS-1, Ras Al-Hamra RH-10, Ras Al-Hadd HD-6, Ras Al-Hadd HD-10, Ras Al-Jinz RJ-2 and Ras Al-Jinz RJ-6, analyzed by XRF.	162
12.21.	Tin concentrations in tools from Wadi Shab GAS-1, Ras Al-Hamra RH-10, Ras Al-Hadd HD-6, Ras Al-Hadd HD-10, Ras Al-Jinz RJ-6, Ras Al-Jinz RJ-2, analyzed by XRF.	162
12.22.	Tin concentrations in ornaments and weapons from Ras Al-Jinz RJ-2, analyzed by XRF.	163
12.23.	Tin concentrations in semi-finished items from Ras Al-Jinz RJ-2, analyzed by XRF.	163

TABLES

3.1.	Main alloys used in antiquity.	13
3.2.	Melting point of ancient metals.	14
3.3.	Main analytical techniques for archaeometallurgical studies.	17
6.1.	ED-XRF quantitative and semi-quantitative analyses of Ras Al-Hadd HD-6 metalworks.	44
6.2.	ED-XRF quantitative and semi-quantitative analyses of Ras Al-Hadd HD-6 metalworks. Semi-quantitative composition: MC = main component; ++++++ → + diminishing content; Tr = trace (modified from Giardino et al. 2007).	45
12.1.	Metal finds from Wadi Shab GAS-1. Alloy normalized concentration. MDL (Minimum Detection Limit) and MRE (Minimum Relative Error) are specified for the elements.	145
12.2.	Metal finds from Ras Al-Hamra RH-10. Alloy normalized concentration. MDL (Minimum Detection Limit) and MRE (Minimum Relative Error) are specified for the elements.	146
12.3.	Metal finds from Ras Al-Hadd HD-6. Alloy normalized concentration. MDL (Minimum Detection Limit) and MRE (Minimum Relative Error) are specified for the elements.	147
12.4.	Metal finds from Ras Al-Hadd HD-10. Alloy normalized concentration. MDL (Minimum Detection Limit) and MRE (Minimum Relative Error) are specified for the elements.	148
12.5.	Metal finds from Ras Al-Jinz RJ-2. Alloy normalized concentration. MDL (Minimum Detection Limit) and MRE (Minimum Relative Error) are specified for the elements.	148
12.6.	Metal finds from Ras Al-Hadd HD-5, Ras Al-Jinz RJ-1, Ras Al-Jinz RJ-4, Ras Al-Jinz RJ-6. Alloy normalized concentration. MDL (Minimum Detection Limit) and MRE (Minimum Relative Error) are specified for the elements.	150
12.7.	Average for element in objects from Wadi Shab GAS-1, Ras Al-Hamra RH-10, Ras Al-Hadd HD-6, Ras Al-Hadd HD-10, Ras Al-Jinz RJ-2 and Ras Al-Jinz RJ-1, analyzed by XRF.	151
12.8.	Crucibles from Ras Al-Jinz RJ-2; int.: inside, ext.: outside. Majority elements (maj): + = 1.00%; ++ = 5%; +++ = 15%; ++++ = 30%. Minority elements (min): + = 0.50%; ++ = 5%; +++ = 10%; ++++ = 15%. Trace elements (trace): + = 0.20%; ++ = 1%; +++ = 5%; ++++ = 10%. MDL (Minimum Detection Limit) and MRE (Minimum Relative Error) are specified for the components.	152
12.9.	Metallic cosmetic amalgam inside a cockle shell from Ras Al-Hadd HD-5. MDL (Minimum Detection Limit) and MRE (Minimum Relative Error) are specified for the elements.	152

Acknowledgments

First of all, I would like to thank His Royal Highness Sayyid Haitham bin Tarik Al Said, Minister of Heritage and Culture of the Sultanate of Oman.

I express my profound gratitude to His Excellency Salim Mohammed Al-Mahrooqi, Undersecretary for Heritage of the Ministry of Heritage and Culture, to His Excellency Hassan Mohammed Ali Al-Lawati, Adviser of His Highness the Minister of Heritage and Culture for Special Projects, and to Mr. Sultan Saif Al-Bakri, Director General for Archaeology of the he Minister of Heritage and Culture, for having put at my disposal the artefacts that I studied.

I thank deeply Prof. Maurizio Tosi, Director of the Italian Archaeological Mission and of the Joint Hadd Project, for involving me in the study of the Omani metals and for the stimulating discussions; he supported this research and provided me with precious information on the early metallurgy of Oman. To him, whose premature disappearance just took place during the final revisions of this volume, I dedicate a grateful and mindful tribute.

I am thankful to Prof. Maurizio Cattani for letting me study the HD-6 metal items and for giving me helpful suggestions. I would like to express my special appreciation to Dr. Alessandra Lazzari for her kind and fundamental collaboration and constant help; to Dr. Francesco Genchi, for sharing his knowledge of the 'Uqdat Al-Bakrah excavation.

Prof. Sobhi Nasir from the Department of Earth Science of the Sultan Qaboos University, who gave me precious help in analyzing the necklace from RJ-2.

I thank Dr. Luigi Tricarico, who made most of the drawings. I am thankful for Dr. Giuseppe Sarcinelli's elaboration of computer graphics. The advice given by Giò Morse in revising the English has been of great help. I am also grateful to Prof. Giovanni Ercole Gigante, to Dr. Giuseppe Guida of the Italian High Institute for Conservation and Restoration (ISCR *Istituto Superiore per la Conservazione e il Restauro*, Rome) and to Dr. Stefano Ridolfi for their invaluable contributions in the scientific analyses of the Omani archaeological materials.

I would like to express my very great appreciation to Dr. Christopher P. Thornton for providing me with the manuscript of his new, precious work on the Bat excavations.

Finally, I would like to express a moving remembrance for Dr. Gert Weisgerber, unforgettable pioneer of the archaeometallurgical research in Oman, who was always generous with advice and information, and who accompanied me on exciting archaeomining surveys.

Foreword

*"Let the land of Magan [bring for you], hard
and resistant copper, diorite stones..."*

These poetic words were written in Mesopotamia, in the ancient city of Ur, more than four thousand years ago, when the last king of Larsa was still ruling. They tell us about a country far away from Sumer, Magan, that was fabulously rich in copper: Magan, the land that the last century researchers have identified with modern Oman.

The prehistoric civilizations of Oman are strongly connected with this red, bright and hard metal, which Omani people learned to exploit and trade since very early times. Copper was an urgent need for the proud states of the Near East. Without this metal, it was impossible to produce the weapons for their armies, the ornaments for their aristocrats, the statues of their gods and kings. It is not possible to understand the history of early Oman without taking into account the precious copper that was mined in its mountains.

This book wants to describe and to analyze the prehistory of the country from a metallurgical point of view, starting from the first appearance of copper artefacts – as an intriguing multifunctional tool found in a fishermen village at Ras Al-Hamra RH-10 – up to the mass recycling activity produced in the large Iron Age workshop recently discovered on the border of the desert of the Rub Al-Khali. To accomplish this goal requires to carefully examine the evidence from archaeological excavations, together with the results of scientific analyses carried out on the metal items that were discovered during these excavations.

The book is divided into chapters that reflect the different chronological periods, which characterized the prehistory of Oman. Metallurgy is in fact a science that cannot be really understood without considering the cultural structure of the societies that produced that technology: metal artefacts and metallurgical residues are in fact an excellent tool for understanding the complex interactions between technology and society.

Chapter 1

A country of environmental diversities

Oman occupies the southeastern corner of the Arabian Peninsula; it is located on the tropic of Cancer, between 25° and 17° latitude north. It has a total area of 309,500 km² and most of central part of the country is covered by a vast gravel desert plain, the Rub Al-Khali (Arabic: "the Empty Quarter"), the largest sand desert in the world. Mountain ranges rise along the north, the Al-Hajar Mountains, while the eastern side of the region is bordered by the sea. Oman has a coastline of almost 3165 km, from the Strait of Hormuz in the north to the borders of the Republic of Yemen in the southwest, overlooking three seas: the Arabian Gulf, the Gulf of Oman and the Arabian Sea.

The country holds a strategically important position, thanks to its geological resources and its location between the sea and the desert, which played a determinant role in Omani civilization and in the spread of metallurgy in the area.

Generally speaking, the Omani climate can be considered to be arid and semi-arid, but it actually differs quite a lot from one region to another. In fact, it is hot and dry in the interior and humid along the coast. The land receives little rainfall except in the mountain areas, where it rains much more. The rainfall changes dramatically within the geographical regions of the country: the average volume of annual rainfall has been estimated at 62 mm, varying from less than 20 mm in the interior desert regions to over 300 mm in the mountain areas (Ministry of Regional Municipalities, Environment and Water Resources 2005). In the north and center of Oman it rains during winter, from November to April, while a seasonal summer monsoon, from June to September, occurs in the southern parts of the country (Dhofar) causing temperature changes. Nevertheless, these atmospheric precipitations allow supplying a large, diffused system of water-bearing layers in the foothill area, which feed the *wadi* basins. Especially in the foothill area, the average rainfall generally has about three peaks that are distributed not only in winter, but also in the summertime (Tosi 1975: 187, fig. 3).

Oman has large amounts of water in aquifers that were replenished when wet climate conditions prevailed, several thousand years ago. Total internal renewable water resources are estimated at 1.4 km³/year. About 1.05 km³ is surface water and 1.3 km³ groundwater, while 0.95 km³ is considered to be the overlap between surface water and groundwater (Frenken 2009: 302, tab. 2).

These peculiar climatic characteristics of the country deeply conditioned the exploitation of the local resources. According to archaeological evidence, they have remained substantially the same, at least since the Copper Age.

The availability of water was determinant to permit a diversified flora and fauna, which made possible to develop the exploitation of the local economic resources since prehistory. The presence of bushes and trees in the *Acacia savannah*, especially along the *wadis*, was relevant for the development of early metallurgy, because these plants could be used with profit as a fuel source to feed the furnaces.

Since the 3rd millennium BC agriculture was developed in Oman thanks to the oasis system; isolated areas of vegetation in an arid environment created by human ingenuity in the surroundings of a water source. Oases provided habitat for plants, animals and even humans (Figure 1.1).

Figure 1.1. Date palms at the Bat oasis (photograph by C. Giardino).

The three geographical provinces of Oman

Geographical criteria allow dividing Oman into three main physiographic provinces: the coast, the mountains and the desert (Frenken 2009: 299). The country is therefore characterized by a large variety of ecosystems that coexist inside its territory; this ecological variability is also increased by the presence of summer monsoons. During prehistory, this variability stimulated different responses by the local communities that developed peculiar specializations in land use and created a mechanism of close interactions and economic integration thanks to the complementary nature of the resources.

The coast and the coastal plain. The country has a long coastline of almost 3165 km, it overlooks three seas: the Arabian Gulf, the Gulf of Oman and the Arabian Sea. The elevation ranges between zero near the sea to 500 meters further inland. The coastal area represents a very favorable biotype for sea life. Many types of fish, mollusks, and coelenterates live along the shore, constituting a rich supply for the human settlements. Fishing villages were established in the lagoons with mangroves along the coast of Northern Oman since the Middle Holocene, which used local shells and stones to produce tools and ornaments. Life conditions should have been particularly profitable; in fact, archaeological research discovered more than one hundred settlements dated from the 4th millennium BC along the coast between Muscat and Dhofar (Cleuziou and Tosi 2007: 69-70).

Figure 1.2. Cliffs at Ras Al-Jinz (photograph by C. Giardino).

These settlements were rather prosperous and had high subsistence levels because they were strongly connected in a network of trades with the other communities spread all over the country. The existence of this network is clearly shown by the presence in one of the fishing sites, at Ras Al-Hamra, of one of the earliest Omani metal objects, a multifunctional tool dating to the end of 4th millennium BC. In the sea villages along the coast near Sur many copper tools related to fishing were in use during the Early Bronze Age, in the context of the Umm an-Nar culture. The copper mines, exploited in the productive centers near the outcrops, regularly reached the settlements along the coast. Metal was therefore in use not only for prestigious items, such as daggers or other weapons, but mainly for everyday purposes, as hooks and needles to repairs fishnets or awls for shell craft, which used the local availability of sea snails as *conus*, *strombus*, *cipraea* and *ostrea*. Also in the later periods, the coastal area remains the best place for us to detect metallurgical innovations. At Ras Al-Jinz (Figure 1.2), the easternmost point of the Arabian Peninsula, the earliest evidence of copper melting was recovered; a much more developed technique in comparison with the simple shaping by hammering that took place at the earlier site of Ras Al-Hadd HD-6 (Figure 1.3).

The mountains. The mountain ranges occupy 15% of the total area of the country. In the southeastern area, the mountains run from Musandam in the north to the Ras Al-Hadd. Here is located the Jebel Al-Akhdar, with a peak that reaches 3000 meters, near the Al-Batinah Plain. Other mountains are in the Dhofar province, with peaks from 1000 to 2500 meters.

Figure 1.3. The coast at Ras Al-Hadd (photograph by C. Giardino).

Figure 1.4. Mountains near Al-Moyassar (Maysar) (photograph by C. Giardino).

Figure 1.5. Rub Al-Khali landscape near 'Uqdat Al-Bakrah (photograph by C. Giardino).

The average annual rainfall in the mountain areas has been estimated at over 300 mm, which is much more than the general average for the whole country (Frenken 2009). The mountain areas are characterized by the presence of copper ore deposits (Figure 1.4). These deposits have been exploited since prehistory, as indicated by the Bronze Age mines that were identified during archaeological investigations. The metal sources from the mountains contributed in a significant way to the early development of Oman, sustaining middle and long distance trades with other communities and countries. According to Sumerian texts dating back to the 3rd millennium BC, copper from the Omani mountains constituted a relevant source for Mesopotamian metallurgy for several centuries.

The desert. The interior regions are placed between the coastal plain and the mountains in the north and the south areas; here the elevations do not exceed 500 meters. This area that covers 82 % of Oman is mainly desert, sand and gravel plains. It includes part of the Rub Al-Khali, also known as the Empty Quarter or the Great Sandy Desert (Figure 1.5). The climate is hot with temperature extremes and is decidedly arid with a low and highly erratic rainfall. The region receives extremely little precipitations, generally 20-35 millimeters of average annual rainfall. Rainfall is generally less than 35 mm per annum, and the daily average relative humidity is about 52% in January and 15% in June-July. Daily maximum temperatures average 47 °C in July and August, reaching peaks of 51 °C. The daily minimum average is 12 °C in January and February, although frosts have been recorded (Mandaville 1986; Franken 2009; Llewellyn-Smith 2014). The sands are predominantly silicates, composed of 80 to 90% of quartz and the remainder feldspar (Mandaville 1986). Dune types range from solitary barchan dunes to extensive longitudinal dunes about 300 km long.

Figure 1.6. Acacia tree in the outer margins of the Rub Al-Khali (photograph by C. Giardino).

The Rub Al-Khali has a very limited floristic diversity. There are only 37 species, of which about 17 live mostly around the outer margins; of these, one or two are endemic species (Mandaville 1986; Norton *et al.* 2009). Vegetation may be described as very diffuse, but fairly evenly distributed sand shrub land, interrupted in some parts by near sterile inter-dune floors. Typical plants are *Calligonum crinitum* on dune slopes, saltbush (*Cornulaca arabica*) and tussocks of sedge (*Cyperus conglomeratus*). Other widespread associates are *Dipterygium glaucum*, *Limeum arabicum* and *Zygophyllum mandavillei*. Trees are absent, except around the outer margins, and are typically *Acacia ehrenbergiana* and *Prosopis cineraria* in drainage lines and pans between dunes (Figure 1.6).

In the past, this area was not only a large border, hard to overcome, but also a relevant network of caravan routes that were ideal for communication and trade, connecting Oman with the rest of Arabia. The camel trains were able to withstand harsh conditions: for centuries, they traveled throughout the desert and often carried lucrative goods, such as fine cloth, beautiful textiles, delicious food, useful and precious metals. The peculiar characteristics of this difficult environment made it an ideal location for some specialized metallurgical activities, where the intense use of fire needed a protected location, as it is possible to observe at the Iron Age industrial area of 'Uqdat Al-Bakrah.

Chapter 2

A geological and mineralogical overview

Hundreds of rich copper ore deposits occur in Oman in the mountains of the northern part of the country, the Al-Hajar Mountains, thanks to its peculiar geological structure. The Al-Hajar Mountains run parallel to the Gulf of Oman, forming an arc more than 700 km long and ca. 150 km wide. Instead, the South has almost no metal ore formations, with the remarkable exception of the large hilly island of Masirah, located about 19 km off the southeastern coast. Several geological units reflect the different phases that characterize the geological history of the country (Figure 2.1). Nevertheless, it is useful to divide these rock units into three main groups to better understand the geologic and metallogenic situation of northern Oman (Coleman and Bailey 1981: 9-28; Michel 1993):

a. The basement autochthonous rocks, represented by Paleozoic and possibly Precambrian meta-sedimentary and meta-igneous rocks. Arabian Shelf Carbonates are resting on the metamorphosed Pre-Permian basement; their rocks consist mainly of limestone, dolomite, and mudstone; their fossils indicate shallow water deposition.

b. The Semail Ophiolite, a thick sequence of rocks that probably represents the ancient oceanic lithosphere; it lies on top of the highest levels of the Hawasina Napples, a unit of deep-water silica rich chert, schist and limestone.

c. Thick shallow-water marine limestone of Late Cretaceous to middle Tertiary age deposited directly on top of both the previous rock units, when they were covered by a transgressive sea that deposited limestone and marl.

From a metallurgical and economical point of view, the most important rock unit in the Oman Mountains is the Semail Ophiolite, which contains in its layers nearly all the copper ore deposits of the country. Semail Ophiolite rest above the Arabian Shelf carbonates. According to Reinhardt (1969), Semail Ophiolite was deposited on the Tethyan Sea during Mesozoic time. The Semail is a classic example of ophiolite; Omani Semail is one of the largest exposures of this kind of rock on Earth. The ophiolite can be divided into five rock units, which are, in sequence, from the base to the top, peridotite, gabbro, diabase, plagiogranites and basalt. Basalt and diabase units are very important from a mineralogical point of view, because they host the metal ore.

Massive sulfide copper deposits are present within the pillow lavas and diabase of the ophiolite sequence. The pillow lavas are a characteristic form assumed by the basalt when it erupts underwater: the contact with water quenches the surface and the lava creates a distinctive pillow shape. Many of the world's best copper deposits are concentrated in similar layers of pillow rocks. The copper mineralization also occurs along faults in the gabbro and peridotite. The metal mineralizations in the intrusive Semail Ophiolite are generally assigned to a massive sulfide type and they are similar to the ores located in the famous Troodos Ophiolite in the island of Cyprus, the main copper source in the Mediterranean basin during the Bronze Age. The Omani mineralization resulted from a hydrothermal activity processes that operated at the end of the submarine cycles of volcanism or during quiescence periods (Michel 1993: 23).

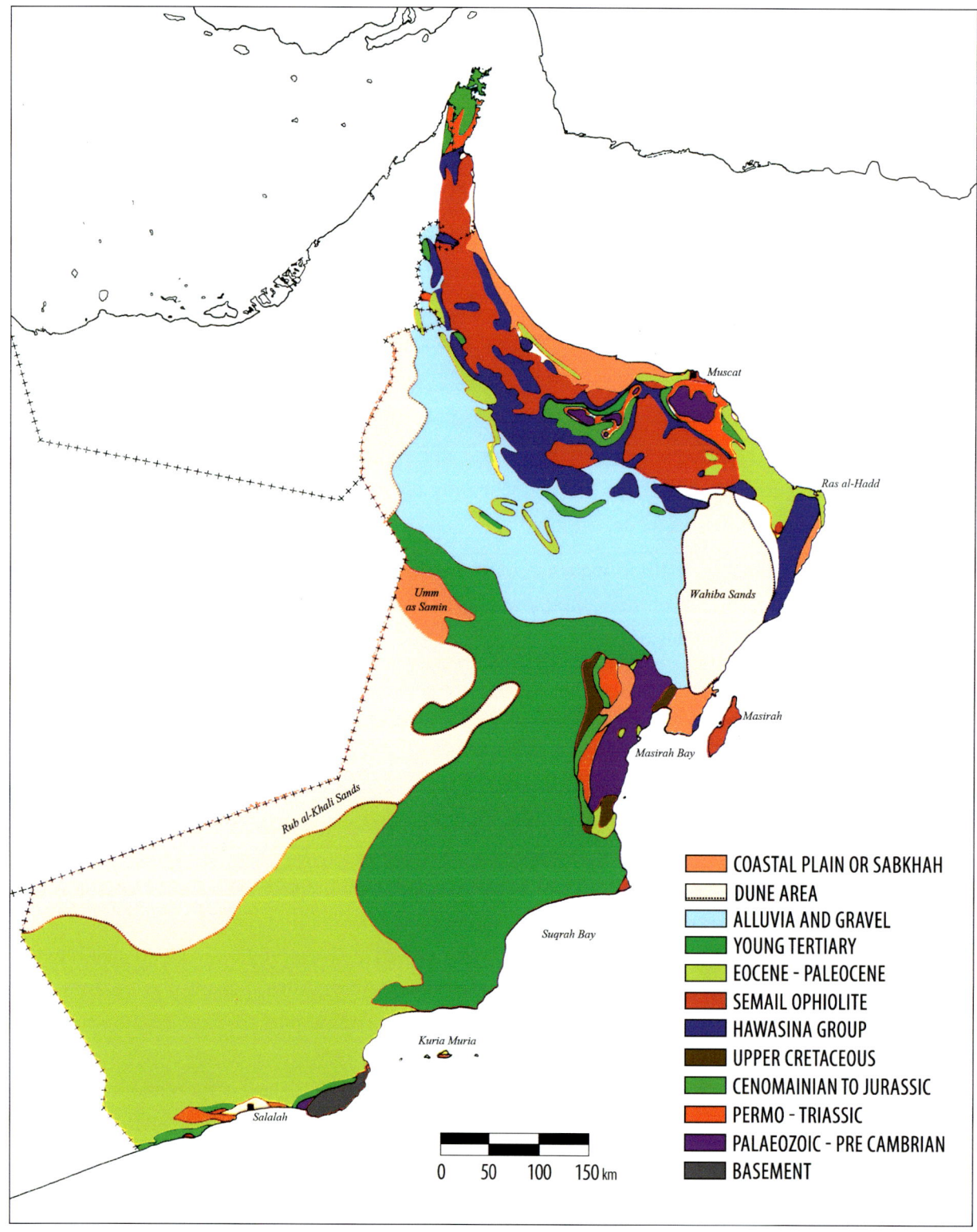

Figure 2.1. Simplified geological map of Oman.

More than 150 copper occurrences are known over the whole Oman Mountain chain (Hauptmann 1985: 25). Semail Ophiolite are spread all over the Al-Hajar Mountains. The southernmost location of this rock formation is around Wadi Musaw, about 60 km south-west of Ras Al-Hadd. Many copper ore deposits are known from the Masirah Island, inside Masirah Ophiolite. None of them is as rich as the deposits from the Semail Ophiolite. Only a small copper ore deposit is known from the southern region of Dhofar, and it is located near Wadi Shatahl in End-Proterozoic–Cambrian rocks belonging to the sedimentary basement.

The Semail Ophiolite and its copper minerals

Semail Ophiolite occupies a vast crescent shaped area that rests upon the northwestern part of Oman for over 500 km. The Ophiolite body is between 50 and 100 km wide and generally between 5 and 15 km thick (Le Métour *et al.* 1995: 87). Copper ore deposits occur in near all the rock sequences of the Semail Ophiolite, but the best ones are near the top of the basaltic pillow lavas. They are accompanied by large amounts of iron sulfide (pyrite) (Figure 2.2).

On the upper and exposed part of the ore deposits the oxidation of the pyrite produced sulfuric acid waters that created colorful, extensively leached gossan outcrops, areas of intensely oxidized, weathered, decomposed rock. In the so-called gossan, the pyrite and primary ore minerals are replaced by iron oxides such as limonite ($FeO(OH) \cdot nH_2O$), goethite ($FeO(OH)$), and jarosite ($KFe^{3+}_3(OH)_6(SO_4)_2$). Therefore, the gossans have remarkable bright colors, red, orange, yellow, brown black: hence, it is very easy to distinguish a gossan on the surface of a territory. In antiquity, the gossans were important guides to buried ore deposits used by prospectors in their quest for metal ores. The Omani gossan zones of the Semail Ophiolite (also called "iron caps") are generally small in surface extent; in plan, they generally have an elliptical shape with major axes of ca. 30-150 m.

The secondary copper minerals of the outcrop are generally located in the oxidation zone of the deposit, under the gossan, but in Oman most of the gossans were still copper bearing in the upper part, therefore the upper weathered zone was mined too (Weisgerber 1980: 115-116). Zones of secondary enrichment are not known from the Semail Ophiolite, probably because they had been found and exploited in the past. Remains of Early Islamic mining activity were discovered at a depth of nearly 100 m.

The most important deposits are presently located at Lasail, Aarja, Bayda near Sohar, Raki and Semdah: they occur inside volcanic rocks (Hauptmann 1985: 25-32); the ore body is mainly pyrite (ca. 95%), containing chalcopyrite. Chalcopyrite ($CuFeS_2$) is the main copper mineral in the veins, but outcrops have generally a rich secondary copper mineralization. Here the prehistoric metallurgists found the most useful ores: azurite ($Cu_3(CO_3)_2(OH)_2$) and malachite ($Cu_2CO_3(OH)_2$), brochantite ($Cu_4SO_4(OH)_6$). Copper carbonates were easier to smelt than chalcopyrite, therefore more useful for the earliest pyrotechnologies.

Two groups can be distinguished between the copper ore deposits of the Semail Ophiolite: the massive sulfide deposits and the veins (Hauptmann *et al.* 1988: 35). The massive sulfide deposits are mainly constituted by pyrite not very rich in copper: its copper content is about 2.0-2.5%; chalcopyrite and bornite (Cu_5FeS_4) both are rare. Irregular mineralized lens and veins occur between gabbros and peridotitic rocks. The main mineral is chalcopyrite, with few amounts of Fe-Co-Ni-A minerals in these deposits, that are generally small, but perfectly useful for a prehistoric exploitation. Outcrops also have a rich secondary copper mineralization with malachite, brochantite and chrysocolla (($CuAl)_2H_2Si_2O_5(OH)_4 \cdot nH_2O$); these ores can reach a copper content up to 30%.

Figure 2.2. Copper outcrop inside the Semail Ophiolite at Bilad Al-Maidin (photograph by C. Giardino).

According to the chemical analyses of copper ore samples from Semail Ophiolite coming from the present mining sites of Lasail, Aarja and other localities, these samples frequently have rather high nickel and cobalt contents, ranging from 5 ppm up to 0.58% of Ni and from 7 ppm up to 0,3% of Co. Arsenic was generally around 0.1-0.37% (Coleman and Bailey 1981: tab. 6; Hauptmann 1985: tab. 2). High nickel content is not typical of all the ores: nickel is generally low in the massive sulfides ore deposits, but it increases in the veins, where it occurs together with cobalt and arsenic. According to the results of analytical studies on prehistoric Omani artefacts, these should probably be the main sources that were exploited in antiquity, as early as the 3rd millennium BC. Scholars observed the association between these ophiolitic copper ore deposits and ancient slag heaps, an evidence that underlines the importance of these smaller deposits for the ancient economy (Weeks 2003: 13).

Other copper ore deposits

Some small copper ore deposits are located in rock units geologically earlier than the Semail Ophiolite. Although manganese is the predominant mineralization in the Hawasina Napples, veins of hematite, galena and copper occur inside the chert of this rock unit that lies underneath the Semail Ophiolite. The main copper ore deposit in the Hawasina Napples is located at Al-Ajal, 60 km west of Muscat, where copper is associated with gold (Weeks 2003: 14).

Some kind of prehistoric exploitation has been hypothesized for the hydrothermal veins located near Bid Bid in the region Ad-Dakhiliyah, about 50 km southwest of the capital. Here copper is mixed with lead, zinc and silver. Ore from Bid Bid is supposed to be the source of copper for a bronze bangle from the Early Iron Age Selme hoard, because it contains an exceptionally high lead content (Hauptmann *et al.* 1988: 47-48). Other small deposits were discovered near Wadi Bani Kharus and near Wadi Samail, which are both in Central Oman. In Southern Oman, the only copper deposit of Dhofar occurs near Wadi Shatahl; it has been discovered in a Proterozoic–Cambrian formation.

Masirah Ophiolite

The island of Masirah was another important source for copper exploitation. Geologists have described the presence of copper mineralization from Masirah since the middle of the 19th century. These ores were inside quartz veins in the form of copper carbonates, like malachite and azurite, in association with hematite (Weeks 2003: 13). The copper ore deposits located inside the Masirah Ophiolite were mined as early as the beginning of 2nd millennium BC: radiocarbon dates from excavations indicate the 18th century BC (Weisgerber 1980: 285, footnote 7).

Chapter 3

Basic elements of metallurgy

The knowledge of metals and the discovery of smelting technologies to produce artefacts from ores are crucial and revolutionary events in human history, comparable to the coming and the spread of agriculture in the Neolithic. The beginning of metallurgy allowed men to use a highly efficient material, much more resistant than bones, wood and even stones. Metals represented a new, durable kind of wealth accumulation: they are not perishable products such as plants and animals, they supported and promoted the concentration of wealth and, therefore, social stratification.

The development of metallurgy fostered many transformations in the economic and socio-cultural structure of the communities, which went far beyond a mere technological aspect. The need for metals promoted long distance, dangerous travels and consequently developed trades and relations between far away peoples of different cultures.

During prehistory, the metallurgists had to move frequently, not only to offer their art and their products, but also to search new areas for mineral exploitation. The Omani sailors and merchants were the protagonists of large maritime trades during the Bronze Age, which involved not only the Arabian Peninsula, but also the Indus Valley. These trades could not be understood without taking into account the urgent request of metal by the rich and powerful states of the Near East.

Properties of metals and alloys

Metal is a material that has peculiar physical, chemical and mechanical properties. A metal is typically hard, opaque, shiny and lustrous, it reflects light, and has good electrical and thermal conductivity. Metals are generally malleable, they can be hammered or pressed permanently out of shape without breaking or cracking; they are fusible, they are able to be melted, and ductile, they are able to be drawn out into a thin wire; they have high density.

An alloy is the result of the mixture of two or more elements in which the main component is a metal. Depending on the number of their components, alloys are defined as binary, ternary, etc. Alloys have metallic characteristics and they have different properties from the individual metals that composed them. Some pure metals are either too soft or brittle for practical use. By combining different metals as alloys, it is possible to modify the properties in order to obtain the desirable characteristics; sometimes alloys have a more desirable color and luster. The most common binary alloys produced in antiquity are bronze (copper + tin), arsenical copper (copper + arsenic), brass (copper + zinc), steel (iron + carbon), electrum (gold + silver) (Table 3.1).

Any pure metal has its own melting point that is the precise temperature at which it changes state from solid to liquid at atmospheric pressure. Among metals commonly in use during prehistory, tin has the lowest melting point (231 °C) while iron has the highest (1535 °C) (Table 3.2).

Copper (Cu)	+ Tin (Sn) → (bronze)
Copper (Cu)	+ Zinc (Zn) → (brass)
Copper (Cu)	+ Arsenic (As) → (arsenical copper)
Copper (Cu)	+ Antimony (Sb)
Copper (Cu)	+ Nickel (Ni)
Copper (Cu)	+ Nickel (Ni) + Arsenic (As)
Copper (Cu)	+ Silver (Ag)
Silver (Ag)	+ Gold (Au) → (electrum)
Gold (Au)	+ Iron (Fe)
Gold (Au)	+ Silver (Ag) + Copper (Cu)
Lead (Pb)	+ Tin (Sn) → (soldering alloys)
Iron (Fe)	+ Carbon (C) → (cast iron / steel)
Iron (Fe)	+ Nickel (Ni) → (meteoritic iron)

Table 3.1. Main alloys used in antiquity.

In the past scholars ascribed the presumed late appearance of iron in common use to its high melting point, assuming that the early furnaces were not able to smelt iron. More recent studies demonstrated that the arrival of iron was as early as the 3rd millennium BC in several contexts in Mesopotamia, Egypt and Anatolia. Surprisingly, not all these early iron objects were made with meteoric iron, but also by iron ores smelting (Waldbaum 1980; Pernicka 1990: 60-62, tab. 7; Giardino 2010: 198, fig. 2). Oman follows the same trend of other Middle Eastern countries: the archaeological excavations recovered a small iron fishhook in an Umm an-Nar context, at Ras Al-Jinz at the site of RJ-2 (DA 10959), that can be ascribed to the second half of the 3rd millennium BC. The melting temperatures of the alloys vary according to the proportions of their components. These temperatures have no relation with the parent metal constituents; in general, the addition of another element lowers the melting temperature, which makes casting easier. Chemical agents, both natural and man-made products, may attack metals, transforming them into chemical compounds: this reaction is known as corrosion. When a metal ore is transformed into metal thanks to metallurgical smelting processes, the latter tends to corrode, that is, to return to its original ore state.

Crystals and grains

The atoms of metals and alloys are arranged in space following an orderly geometric pattern, the crystal lattice. Therefore, a group of atoms forms a crystal, while a group of crystals forms a grain. This structure can be observed with an optical microscope using a magnification around 200 X. A micro-history is recorded in the internal structure of a metal, giving us a basic determination of the techniques used by the ancient metalworkers, through the careful study of the objects that they produced. Metal crystals can be deformed, distorting their outline. The effects of distortion can be removed by heating the metal to a suitable high temperature below the melting point (annealing): this heating causes new unstrained crystals to grow (Smith 1981: 70-71). Generally, the annealing temperature for copper alloys and iron is around 500-800 °C.

Cast and worked metals have two basically distinct types of microstructures (Scott 1991: 5-10). When a molten, liquid metal cools down, it forms inside it crystals and grains. The size and the form of the grains vary according to the mechanical and thermal treatments. During casting the metal generally assumes a dendritic structure, a characteristic tree-like structure (the term dendrite comes from the Greek word "dendron", which means tree) (Figure 3.1).

Almost all ancient metals are impure, i.e. they are deliberate or involuntary alloys of two or more metals. Crystallized alloys are composed by millions of dendritic crystals all stuck together, because each one of the constituents of the alloy has a different melting point. The faster the metal cools, the smaller the dendrites. Occasionally a metal is not an alloy: in this case, it produces an equiaxial hexagonal grain structure (Figure 3.2).

Metal	Melting point
Mercury (Hg)	- 38.8 °C
Tin (Sn)	232 °C
Lead (Pb)	327 °C
Zinc (Zn)	420 °C
Antimony (Sb)	630 °C
Silver (Ag)	960 °C
Gold (Au)	1063 °C
Copper (Cu)	1083 °C
Iron (Fe)	1523 °C
Platinum (Pt)	1774 °C

Table 3.2. Melting point of ancient metals.

Metals can be shaped by working. There are many working techniques, as hammering, turning, raising, sinking, drawing: the artisan has to choose among the different techniques in order to obtain the wished result. All the workings deform the microstructure. Hammering flats grains and dendrites, until they become brittle and the metal must be annealed to restore malleability and ductility. This process produces a typical recrystallization that creates twinned grains (Figure 3.3). Generally, the objects are shaped by many cycles of hammering and annealing. Hammering and annealing leave peculiar marks on the structure of the object. In this way, it is possible to detect the working undergone by the metal by looking at its crystalline structure: the study of the metallographic micro-structure contributes to reveal the operation chain to produce ancient objects (Figure 3.4).

Analytical techniques for archaeometallurgical studies

Many different analytical techniques are used in order to characterize old metal finds (Table 3.3). Analytical tools, such as Optical and Scanning Electron Microscope (SEM), X-Ray Fluorescence (XRF), X-Ray Diffraction (XRD) and Inductively Coupled Plasma Mass Spectrometry (ICP-MS), are widely employed in the study of ancient metals. However, the most useful technique to determine the ore source of ancient metal items is the Lead Isotope Analysis.

The Scanning Electron Microscope produces images of a sample by scanning it with an electron beam. This electron beam interacts with atoms in the sample, producing various signals. They are detected in order to have meaningful information about the elementary composition of the sample and the micro-topographic location of the different elements on its surface (Figure 3.5). SEM magnification has a range from about 10 to 500,000 times. X-Ray Diffraction is an analytical technique based on X-Ray beams that reveals information about the crystal structure and the chemical composition (but not the elementary composition) of crystalline materials like metals. Inductively Coupled Plasma Mass Spectrometry (ICP-MS), Inductive Coupled Plasma with Optical Emission Spectrometry (ICP-OES), Atomic Absorption Spectroscopy (AAS) are some of the most common techniques employed in archaeometallurgical studies in order to analyze trace elements.

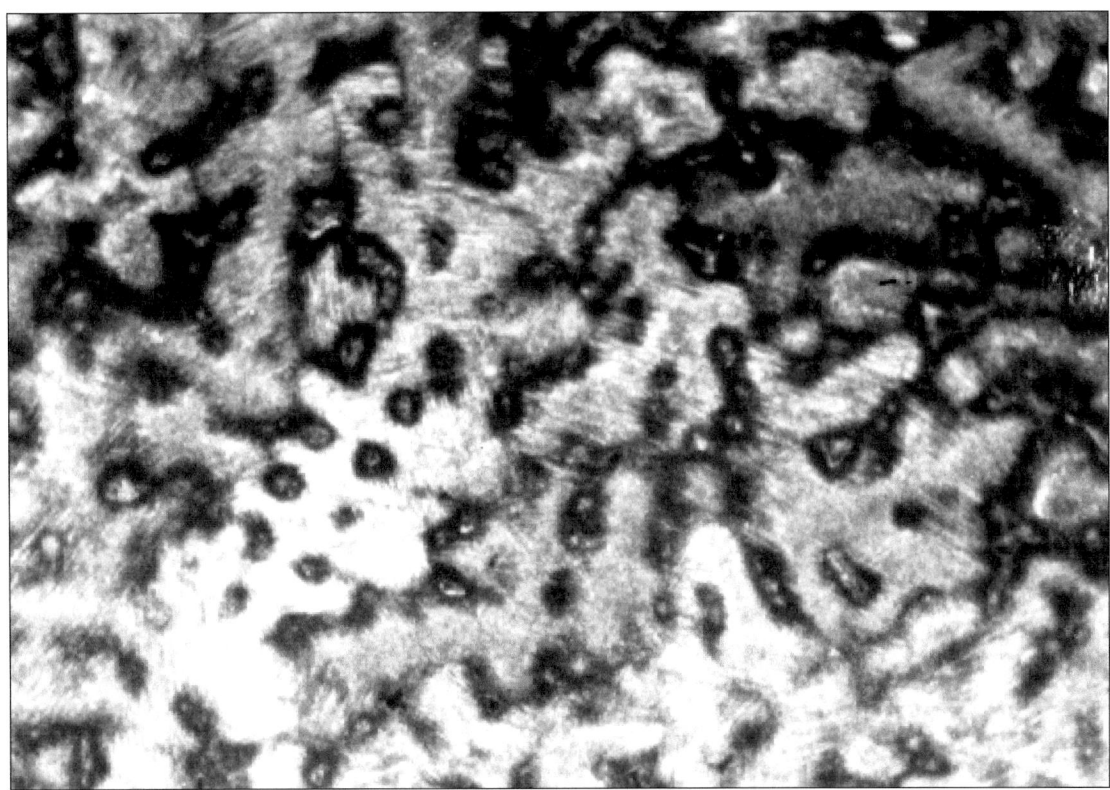

Figure 3.1. Photomicrograph of a copper-based object after casting showing a dendritic structure (after etching) (photograph by C. Giardino).

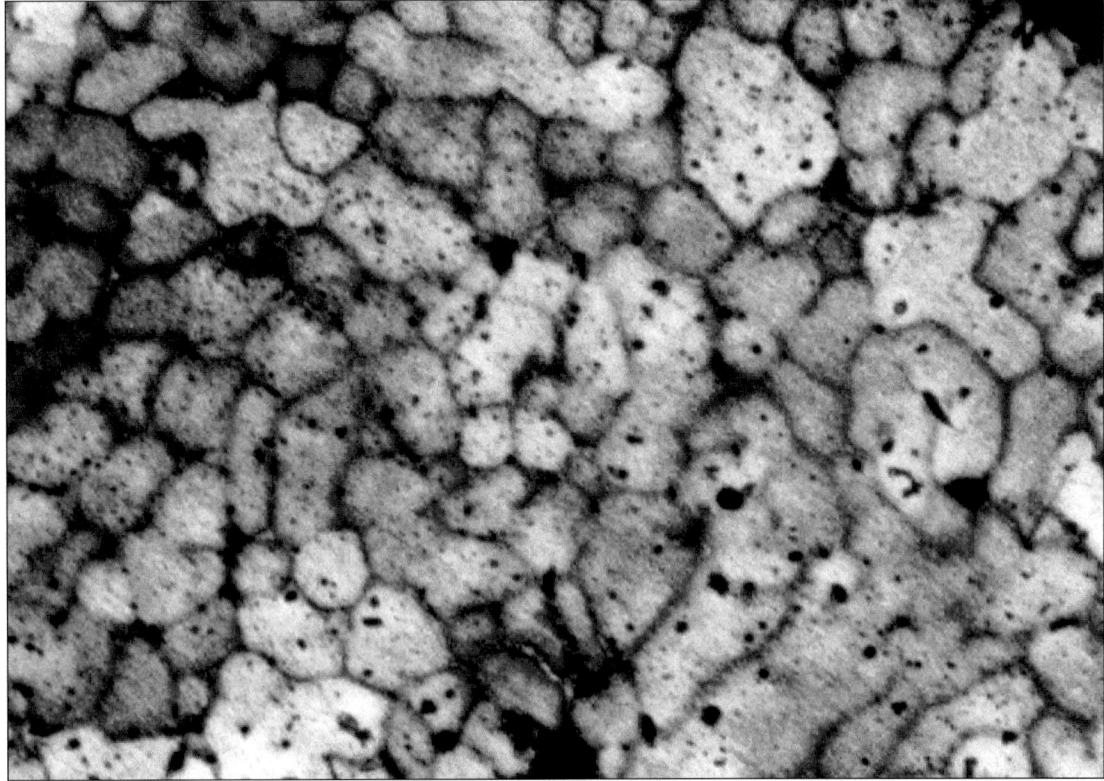

Figure 3.2. Photomicrograph of a copper object after casting showing an equiaxial hexagonal grain structure (after etching) (photograph by C. Giardino).

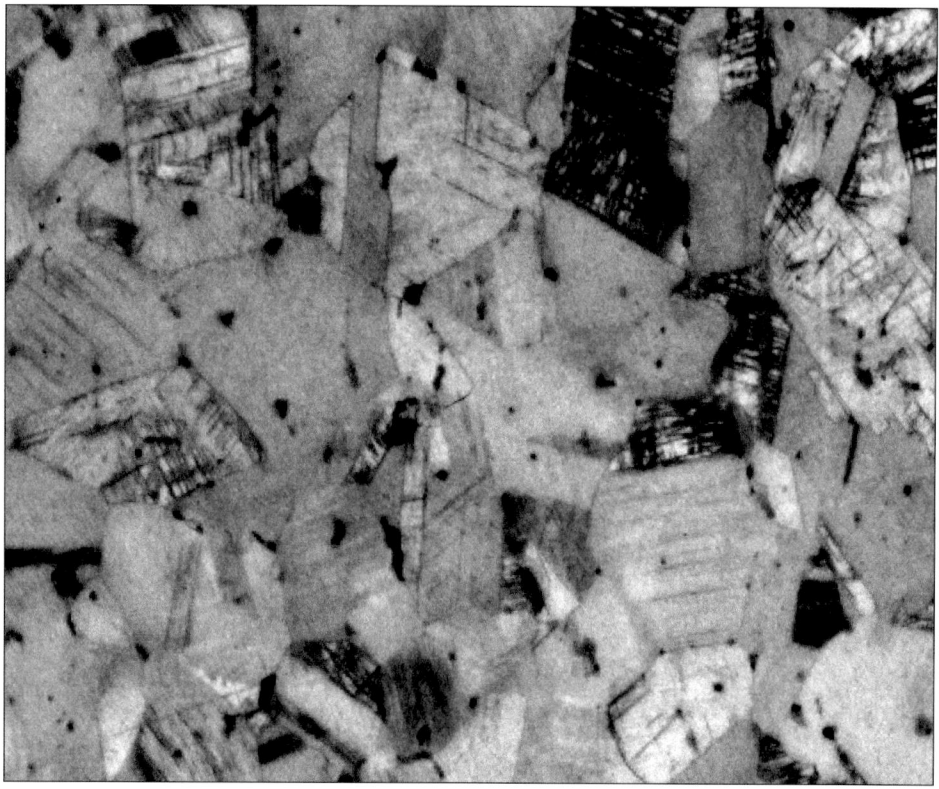

Figure 3.3. Photomicrograph of a copper object after hammering and annealing showing twinned grains (after etching) (photograph by C. Giardino).

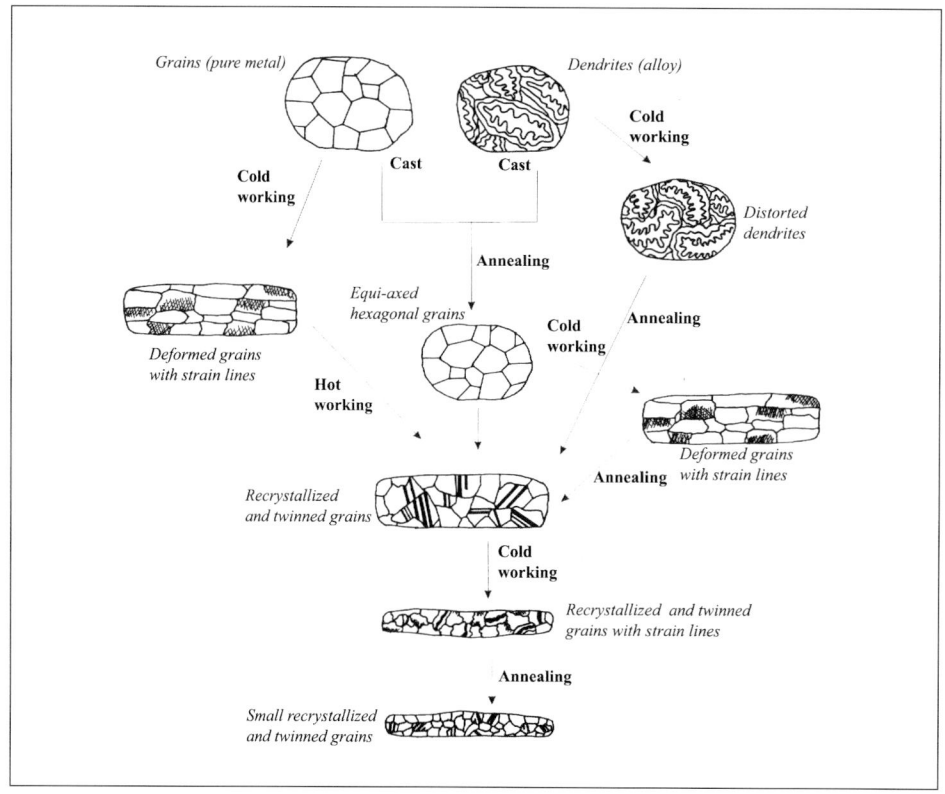

Figure 3.4. Microstructural changes in a metal submitted to mechanical and thermal treatments (after Giardino 2010).

Analysis	Acronym	Characteristics	Results	Destructive Non-destructive	Studies
X-Ray Fluorescence	XRF	Quick and relatively cheap	Elementary quantitative analyses	Non Destructive	Studies of alloys; studies of the content of crucibles
Atomic Absorption Spectrometry	AAS		Elementary quantitative analyses (with trace elements)	Invasive: needs sampling	Studies of alloys
Inductive Coupled Plasma with Optical Emission Spectrometry	ICP-OES		Elementary quantitative analyses (with trace elements)	Invasive: the analysis needs sampling	Studies of alloys
X-Ray Diffractometry	XRD	Studies of patinas, mostly for restoration	Chemical compounds analyses	Invasive: the analysis needs sampling	Studies of patinas, mostly for restoration
Optical Microscopy	OM		Microstructure	Non Destructive / Invasive: the analysis may need sampling	Studies of working techniques
Scanning Electron Microscopy	SEM		Elementary quantitative analyses / Microstructure	Non Destructive / Invasive: the analysis may need sampling	Studies of alloys / Studies of working techniques
Lead Isotope Analysis Mass Spectrometry	LIA-MS		Lead isotopes ratio	Invasive: the analysis needs sampling	Provenance of the ore

Table 3.3. Main analytical techniques for archaeometallurgical studies.

They can detect even very small quantities of an element in a sample. For example, Inductively Coupled Plasma Mass Spectrometry is capable of detecting metals at concentrations as low as one part in 10^{15}. Lead Isotope Analysis is at present the most useful and successful method for establishing the geographical provenance of lead present in ancient metals. It analyzes the variations in lead isotope ratios contained in copper that are measured very accurately using mass spectrometry. It separates the different isotopes of lead based on their mass-to-charge ratio. It is based on comparisons of three lead isotope ratios: 206Pb, 207Pb and 208Pb (Stos-Gale et al. 1995; Gale and Stos-Gale 2000). A significant role is now played by X-Ray Fluorescence (XRF), a technique that recent improvements in instrumentation has made very reliable and that was widely used to analyze the prehistoric archaeological items from Oman by the author (Figure 3.6).

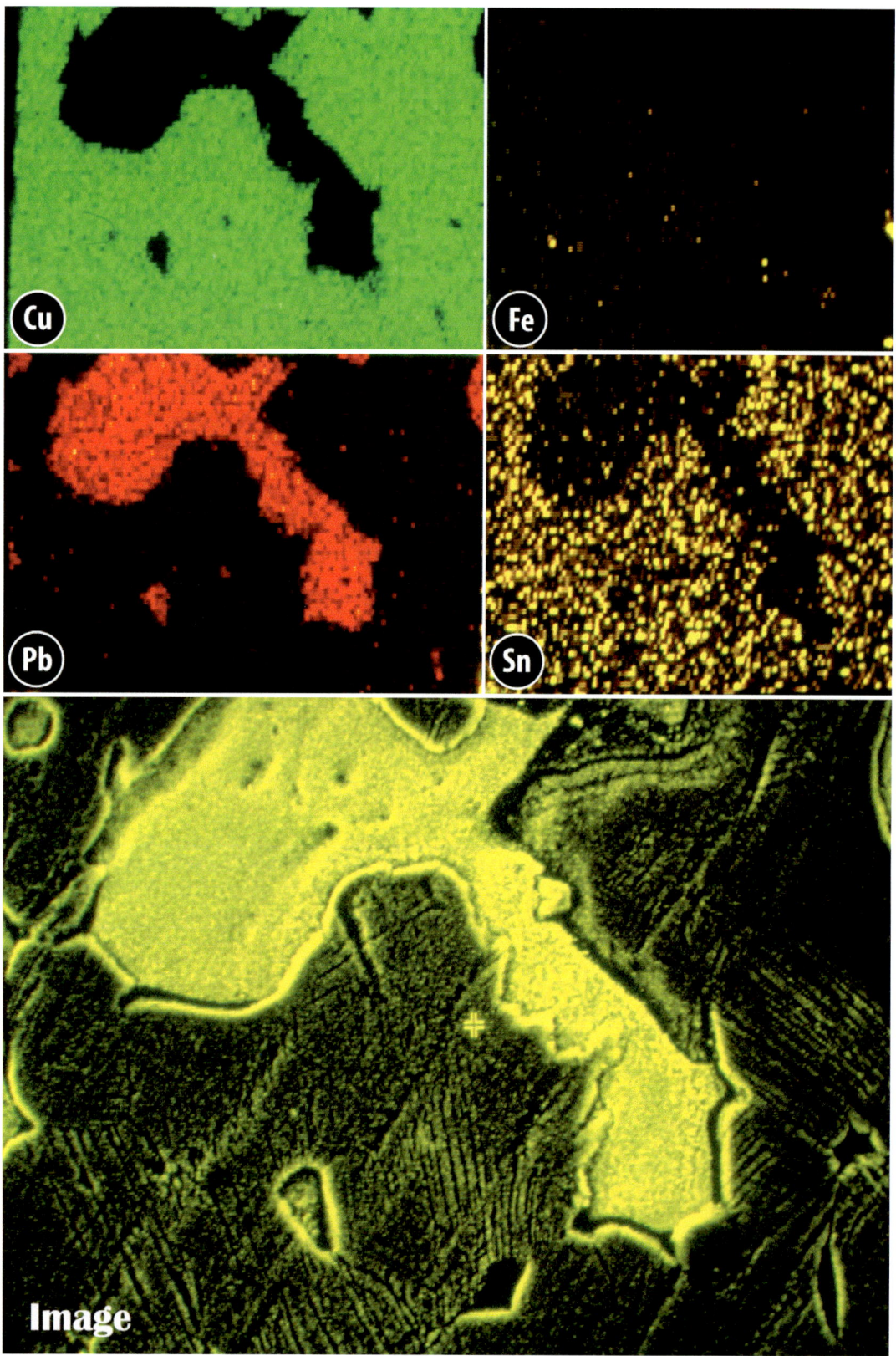

Figure 3.5. SEM image with elementary micro mapping in a micro-structural feature.

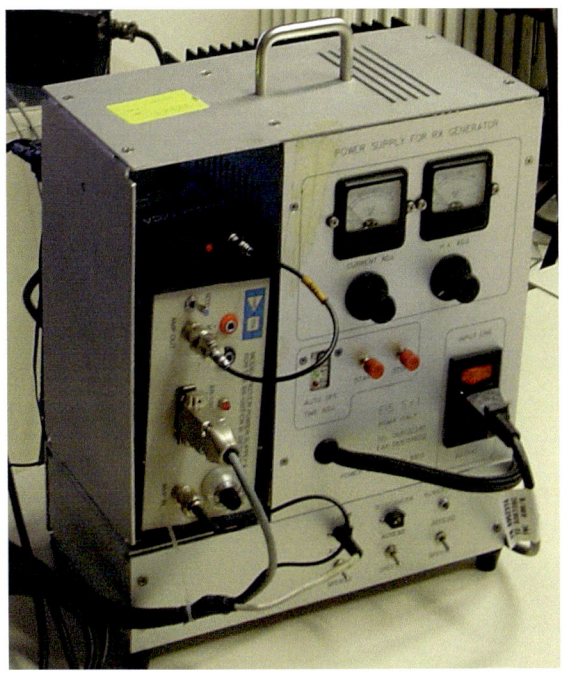

Figure 3.6. X-Ray Fluorescence portable equipment.

The current portable detectors reach resolutions that are at the limits of the detector natural resolution and that were obtained, in the past, only with complex and expensive equipment. The easy portability of the instrument makes possible avoiding the displacement of archaeological finds, which are analyzed in situ where stored. Without denying the inherent limitations of this analytical technique, such as its low penetration power, the choice of XRF to study many archaeological contexts is its portability. Moreover, XRF is non-invasive, non-destructive, low cost, fast and it has good sensitivity for all elements of archaeometallurgical interest. These characteristics make this technique ideal to analyze large amounts of ancient artefacts, where the main interest is the composition of major and minor elements or in studying the composition of the corrosion layers for restoration purposes.

Chapter 4

Copper for Sumer

Mesopotamia is a large alluvial plain, created by two main rivers, the Tigris and the Euphrates. Because of its geological origin, it has no metal ore resources. Nevertheless, this area has very early evidence of copper metallurgy: at Tell Magzaliya, a site in northern Iraq, an awl found in a Neolithic pre-ceramic context is dated 8000-7500 BC. Obviously, metal was imported from other countries, in this case, probably from central Iran (Ryndina and Yakhontova 1985: 156-165, fig. 1; Pernicka 1990: 43, tab. 4); native copper, like other valued goods such as turquoise and obsidian, was traded in small quantities.

It is during the late mid 5th and the 4th millennium BC that the production and use of metal objects changes dramatically, with the emerging of a large demand for metals; this phenomenon is connected with the emergence of a complex society (Heskel 1983). During the Uruk and Akkadian periods (4th – end of 3rd millennium BC), Mesopotamia developed an impressive metal industry, with a large use of tin-copper alloys for the production of weapons, tools and other implements. The need for greater amounts of raw material, to satisfy the increasing request of the metal industry, was a driving force for the economy of Mesopotamia and other countries reached by long-distance trade. Mesopotamian Bronze Age cuneiform texts give us relevant information about these trades, recording the toponyms for some copper provenances. In particular, some documents quote the name of the lands of Dilmun and Magan. Dilmun appears in Sumerian mythological compositions and in official and economic texts; the place is associated with Sumer, as a trading crossroad that imported raw materials vital to Sumer, the center of the civilized world.

The earliest reference to Dilmun is in proto-literate texts from Ur, dating around 3100 BC. Hundreds of kilograms of copper from Dilmun reached Mesopotamia during the 3rd millennium BC, but the trade reached its peak in the early 2nd millennium BC, in the Isin-Larsa and Old Babylonian periods (André-Salvini 2000; Potts 1990; Weeks 2008: 89). The boats from Dilmun arrived at the powerful international ports of Ur-Nanshe of Lagash, Sargon of Akkad, Gudea of Lagash, Ur of the Larsa dynasty throughout more than eight centuries. These boats were regularly bringing to Mesopotamia fabulous riches, such as metals, wood and precious stones (Figure 4.1).

The land of Dilmun is frequently associated in the Sumerian texts with two other driving forces, the countries of Meluhha and Magan. The three countries appear to be strongly connected in a southern Mesopotamian clay cuneiform tablet dated to the beginning of the 2nd millennium BC that reports the Sumerian creation myth known as "Enki and the World Order" (Enki was the Sumerian deity of creation, water, intelligence and crafts) (André-Salvini 2000: 28):

"...The lands of Magan and and Dilmun come to me, Enki, let the anchorage be prepared to receive the boats from Dilmun; let the boats of Magan reach the horizon; let the great (cargo) boats of Meluhha transport the gold and silver..."

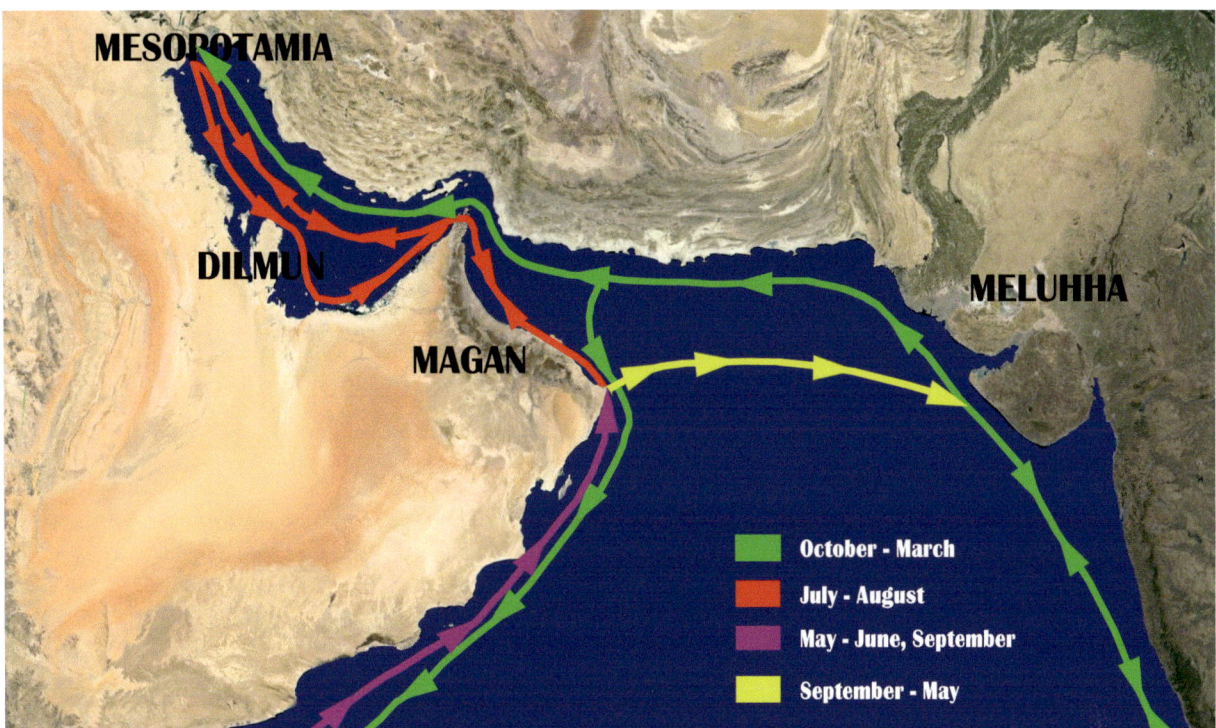

Figure 4.1. Trade routes during the 3rd millennium BC, according to the season winds.

A long lyrical hymn found at Ur, dating at the last king of Larsa, known as "Enki and Ninhursag", specifies the kind of products that arrived at Dilmun from different countries (André-Salvini 2000: 29):

"...*Let the land of Meluhha* [bring for you], *on great ships, desirable and precious carnelian, trees of Magan* [...] *Let the land of Magan* [bring for you], *hard and resistant copper, diorite stones...*"

According to Sumerian literature, Magan and Meluhha were willing, peaceful and indispensable trading partners of Dilmun. Dilmun is identified with Bahrain, a group of thirty islands and islets located close to the eastern coast of the Arabian Peninsula. The archipelago is situated halfway from the Strait of Hormuz and the mouths of the Tigris and Euphrates, in a strategic position on the Gulf, an ideal condition to be a crossroad for maritime routes.

The importance of this land for Sumerian trade is testified by a large quantity of archaeological finds of Mesopotamian origin discovered on the eastern coast of Saudi Arabia, opposite Bahrain, which increased at the beginning of the 3rd millennium BC. The archaeological evidence demonstrates that ancient Dilmun was not only an insular place, but it also had a continental aspect, even if the consolidated sea trade with southern Mesopotamia quickly increased the role of the islands (Cleuziou 2000).

The oldest texts known about Dilmun date around 3300 BC; also in these very early texts its name is associated with copper, even though in Bahrain there are no metal ore deposits, because the geological nature of the country is sedimentary. Therefore, copper had to arrive at Dilmun from other places, through a network of trades. It was ascertained that the end of the 4th millennium BC is just the time when Oman began its mining activities (Cleuziou 2000: 25).

Copper was traded in plano-convex ingots, a shape that was very common in the Near East during the Bronze Age. Some ingots were found in Bahrain, as items discovered at An-Nasiriyah, on the western part of the main island (Lombard 2000: 78). Similar ingots were recovered from Omani Bronze Age sites, like at Al-Moyassar in the Wadi Samah, a smelting site near the mines (previously spelled Maysar). Sumerian administrative documents refer to the arrival in Mesopotamia of copper from Dilmun. One such document is a cuneiform tablet dating to 2300 BC from Girsu, that testifies that the merchant Ur-Enki brought 85 kg of copper to Lagash, to Dimtu, king Enentarzid's wife (Crawford 2000: 77).

The relevance of copper for the economy of Dilmun is also demonstrated by finds from the largest archaeological site in Bahrain, the tell of Qal'at Al-Bahrain, on the north coast of the main island. Here in the earliest occupation layers of the settlement, dating back to ca. 2200 BC, a large metallurgical workshop was recovered; it proves the abundance of metal in the country. The workshop covered an area of 15 x 35 m; inside many open molds were uncovered, together with crucibles. Most of them had an estimated volume of 300-800 cm^3, but some could contain around 4000 cm^3 of molten metal, suggesting a large-scale production, ruled by full time specialists. The analyses carried out on copper coming from one of the crucibles show that it is compatible with copper from Oman.

The connection between Oman and Qal'at Al-Bahrain does not only consist in the large-scale transportation of metal. The pottery found at the site is mostly local, but there are also vessels imported from Mesopotamia and Oman. Moreover, steatite bowls suggest additional links with the Umm an-Nar communities of Oman (Højlund 2000: 59-60). Other Omani stone vessels made with soft chlorite and steatite were recovered in the Saar and Al-Hajar necropolises; some of them are decorated with a dotted double circle pattern. The shape of the stone vases from the Dilmun graves are typical of the Omani Bronze Age cultures of Umm an-Nar and Wadi Suq.

Clay vessels of Omani tradition were also recovered from Dilmun tombs, from Medinat Hamad and the Saar cemeteries. Their typology links them to the Umm an-Nar decorated pots, dating back to 2300 BC. The petrographic analyses carried out on several of these items from Dilmun confirmed that they were made of clay coming from the Omani foothill region. Nevertheless, some of the vessels were not just imports, but they were locally made in Bahrain, following Omani traditions. As demonstrated by archaeometric analyses, the Dilmun potters copied the foreign shapes and decorations, with local Bahrainian clay (Lombard 2000; Crawford 2000). The presence of this very peculiar, specialized production is not only a proof of the great ability of the Dilmun potters, but it is also an indication of the presence of Omani merchants in Bahrain who followed their products. They were somehow integrated into the local communities: these traders preferred to use a pottery that followed their own style and tradition.

Dilmun was in fact a large trading port, a crossroad that linked the Near East to the Gulf and the Indian subcontinent, connecting Mesopotamia, Oman, Iran and the Indus Valley, up to Badakhshan in Afghanistan. As always, this great emporium hosted different groups of people coming from various places, bringing with them not only goods, but also ideas, customs, traditions. The local production of pottery with foreign styles reflects the multi-ethnical character of the country. Dilmun was the place where foreign boats arrived to trade their merchandise and to resupply fresh water and provisions before starting their long travel to Mesopotamia.

The Sumerian documents testify the presence of vessels from Magan in the Mesopotamian ports. A cuneiform text dating to Sargon of Akkad (2334-2279 BC) states that (*Sargon b 2 = RIM E2.1.1.11*, Del Monte 2004: 13):

"...Sargon, king of Kiš: he achieved victory in 34 fights and has destroyed the walls (of the defeated cities) until the shore of the sea; he got ships of Meluhha, ships of Magan and ships of Dilmun to land the docks of Akkad..."

Another text dating to the Third Dynasty of Ur says (*Ur-Nammu 26 Clay Cone = RIM E3/2.1.1.17*, Del Monte 2004: 24):

"...Ur-Nammu, the powerful, king of Ur, king of Sumer and Akkad, who built the temple of Nanna... gave back in Nanna's hands the ships of Magan..."

This means that Magan had to have its own fleet to transport its copper, and that its watercrafts were well known to the Sumerians. A cuneiform administrative document from Girsu – describing materials for boat building – mentions (*CT-7-31*, Heimpel 1987: n. 39):

"...3170 gur (= 951 m3) of asphalt for the coating of Magan type boat..."

This last document is a clear indication that the boats of Magan had a characteristic aspect, large and black: that is why the Sumerian scribe had associated the vessels from this country with bitumen-coated boats (Figure 4.2). Fragments of this coating were discovered at Ras Al-Jinz during excavations; impressed slabs of a bitumen compound were recovered in buildings of that site, in a context dating back to 2500-2200 BC (Cleuziou and Tosi 1994). Bitumen had waterproof and antifouling function for the hull (Baldacci 2011: 15). It has been estimated that the actual cargo capacity was around 15 metric tons (Vosmer 2008: 230-231). The ships followed the coast, following seasonal routes according to the winds (Vosmer 2007: 208-209).

The prosperity of Dilmun was mainly based on copper trade. The situation changed dramatically around 1700 BC, when Ur, the main export port, was destroyed during a revolt against Hammurabi's son, drying up the market of south Mesopotamia. Furthermore, copper from the Arabian Peninsula became less attractive because of the availability of new, cheaper sources closer to Mesopotamia, in Anatolia and Cyprus (Crawford 2000: 76).

Two trade partners of Dilmun were recorded in the literature, Meluhha and Magan. The first one is identified with the Indus Valley, an area that the archaeologists associate with the Indus (or Harappan) Civilization, including southern Baluchistan and Gujarat. According to most scholars, in the 3rd millennium BC the term Magan included the whole of Oman or parts of this country (Cleuziou and Tosi 1994: 746; Possehl 1996: 136-138; André-Salvini 2000: 28). Indus Valley entrepreneurs therefore frequented Magan. Strong connections between Oman, the Indus Valley and Bahrain are attested in the archaeological record. Dilmun was a port of transit for Meluhha and Magan ships, but the two areas had a long tradition of interchange, as demonstrated by the archaeological evidence found in Indian and Omani sites, at Ras Al-Jinz, Hili and Mohenjo-Daro (Cleuziou and Tosi 2004: 172, 181, 184, figs. 176, 189, 193). Merchants from Magan were probably the intermediaries among the great economic circuits of Mesopotamia and the Indus Valley, thanks to the technical capability of the Omani fishermen, who accumulated centuries of experience travelling throughout the Indian Ocean to connect Arabia with India (Cleuziou and Tosi 2004: 178).

Figure 4.2. Reconstruction of an Early Bronze Age Magan boat (photograph by Helen Kirkbride, courtesy Oman Maritime Traditional Boat Centre).

In the Sumerian literature, Magan is always associated with copper; therefore, this country had to be the main copper source for the Dilmun traffic directed to southern Mesopotamia. Unfortunately, Mesopotamian texts are not so explicit in geographically locating Magan as they were for Dilmun; still, they affirm that it lay south of Sumer. A geographical tablet of Sargon of Akkad (ca. 2334 - 2279 BC) states (Peake 1928):

"...Dilmun (and) Magan, countries beyond the Lower Sea and the countries from the rising to the sitting of the sun which Sargon the... king conquered with his hand..."

Looking for Magan

Since the end of 19th century there had been speculation among the scholars about the precise location of Magan; most of them placed Magan in the eastern part of the Arabian Peninsula; nevertheless, another school of thought – basically philologists and assyriologists - would place it in different areas: in Africa, in Egypt, in Nubia, in Ethiopia or in Sudan (Rogers 1895: 7, 13; Jastrow 1915: 136; Perry 1937: 60-61; Kramer 1970: 61, 139, 276-284).

The theories of these scholars, however, were based mainly on late Assyrian texts (in connection with the king's campaigns to conquer Egypt in the 8th – 7th century BC), which are more than one thousand years later than the early Akkadian sources. It is hard to locate Magan in Africa following the Ur tablets and the archaeological evidence. There were no imports from Egypt to Mesopotamia – nor vice-versa – dating to 2000 BC (Bibby 1996: 160).

Magan had to be rich not only in copper ore deposits, but it should also have had large amounts of diorite of remarkable quality that was quarried in its mountains. Cuneiform tablets refer that this high-valued, gray igneous rock was actively exported to Mesopotamia; many statues of Sumerian rulers were made with this precious material (Winter 2010: 123).

On the base of a statue from Susa, the inscription says that king Naram-Sin of Akkad (ca. 2254 – 2218 BC) cut diorite blocks from the mountains of Magan during a military campaign against this country. Then he brought them to Akkad to have a triumphal statue of himself carved (*Naram-Sin Statue A from Susa = RIME 2 E2.1.4.13*, Del Monte 2004: 15):

"*...Naram-Sin, the powerful, king of the four parts of the world, victorious in nine battles in a single year. After he winned these battles he abducted their three kings in chains in front of Enlil [...] He subdued Magan and captured Manium, the lord of Magan, took blocks of diorite on their mountains, carried (them) in his city Akkad, made a statue of himself...*"

Another inscription refers that the same was done one century later by Gudea of Lagash (Figure 4.3) (*Gudea Statue B V 21-VII 20*, Del Monte 2004: 20):

"*...[Gudea, Lord of Lagash] carried diorite from Magan, made this statue with it, called it 'I have built his temple to my king; the life is my part' and introduced it for him in Eninnu (the temple of the high god Enlil in Lagash)...*"

Petrographic analyses on some samples of old Akkadian stones – including the Gudea's statue – demonstrated that the rock is an olivine-gabbro; the statue is an equigranular quartz diorite. Olivine-gabbro is available in Oman, but the diorite that exists in that country is not useful for statue manufacturing. Both these rocks occur in the region of Bandar Abbas, on the southern coast of Iran, on the shore of the Gulf opposite Mesopotamia (Heimpel 1982; Glassner 1989: 186). Some scholars think therefore that Magan was, for the people of Mesopotamia, a faraway country, beyond Dilmun, located on both sides of the Straits of Hormuz (Possehl 1996: 135).

Furthermore, Magan had to be a mountainous country: in fact, the texts recorded the "mountains of Magan" (Weisgerber 1983: 269). The presence of a structured socio-political organization in Magan is suggested by the presence of large scale copper production and exportation, which needed some kind of hierarchy. In fact, some Sumerian texts deal with the "lords" or "kings" of Magan, generally in a context of Sumerian military actions, like the battles of Naram-Sin (*Chronicle of Early Kings*, Del Monte 2004: 49):

"*...[Naram-Sin, son of Sargon] went against Magan and [captured] Mannudannu, the king of Magan...*"

Figure 4.3. Diorite statue of Gudea of Lagash (courtesy Musée du Louvre).

Archaeological and archaeometallurgical research

Aside from philological research, based on Sumerian written sources, archaeological studies on where Magan might be located started at the beginnings of the last century, with the research of Harold Peake (Peake 1928). He analyzed 20 copper-based specimens of archaeological objects from Sumerian settlements and graves, and from Bronze Age tombs in Bahrain, together with several samples of copper ores from Anatolia, Iran, Cyprus, the Sinai Peninsula and Arabia. Peake observed that most of Sumerian objects had considerable and unusual high nickel content, ranging from 0.05% up to 3.34%. No trace of nickel could be found in the copper minerals that were analyzed, with the only exception of the ore from Oman. It was collected from some old copper workings at Jabal Al-Ma'adan, in Wadi Ahin, around Sohar. According to these results, Peake stated that most probably Oman could have been the copper source for the early people of Mesopotamia, and that this country could be identified with the ancient Magan of Sumerian literature. The archaeological excavations carried out in the 1950s at Umm an-Nar – a coral island in the United Arab Emirates near Abu Dhabi, adjacent to the western coast of Oman Peninsula – gave a new input to the Magan question. Here metal workshops together with sophisticated materials such as imports from Iran, the Indus Valley and northern Syria indicated that the coastal people had, in the second half of the 3rd millennium BC, a high degree of affluence and they actively participated in the trade circuit between India and Mesopotamia (Tosi 1989: 139-140).

A new interest in the location of Magan occurred once again about half a century later, when evidence from excavations in Bahrain proposed the question anew (Bibby 1970). Unfortunately, the Sultanate of Oman remained politically isolated until 1970, when His Majesty Qaboos bin Said Al Said rose to the throne. The proper archaeological research in Oman began from that date. Geological and archaeological research in Oman has recovered significant copper deposits and more than 150 medieval Islamic smelting sites since the 1970s. During the Seventies, the investigations carried out by the Harvard Archaeological Survey in Oman found evidence of Bronze Age copper production at the sites of Samad 5, Zahir 2/3 and Batin 1 (Hastings, Humphries and Meadow 1975). The Harvard team collected fragments of crucibles and other casting waste in various sites dating to the 3rd millennium BC at Wadi Andam. They discovered some crucible fragments also in the proto-historic settlement of Wadi Ibra 2, dating back to the 3rd millennium BC; they were conical shaped crucibles 12-15 cm high, whose capacity was about 500 cm^3.

Copper prospecting work carried out by Prospection Ltd. (today Oman Mining Company L.L.C.) identified many ancient mines and large slag heaps, but none could certainly be ascribed to a prehistoric exploitation. Some mines were observed and examined, Lasail and Arja on Wadi Jizzi, Masjad on the Luzaq: they were all medieval mines, but the smelting slag found nearby showed a technology very close to that used during the Bronze Age (Tosi 1975: 196-202). These very promising perspectives were confirmed by the large-scale research carried out since 1977 by the German team from the Deutsches Bergbau Museum of Bochum. They discovered that nearly all the slag waste was of early Islamic origin, which frequently obliterated prehistoric features, but they also recognized some 4000 years old smelting traces. The Germans concentrated their research in the Al-Moyassar Valley along Wadi Samad, in central-eastern Oman, where evidence of Bronze Age copper production was identified (Weisgerber 1977, 1983; Hauptmann and Weisgerber 1981). The excavations of the Deutsches Bergbau Museum identified many Magan-period (2500-2000 BC) slag heaps and prehistoric remains of mining and smelting at this site, where a hoard of plano-convex copper ingots, the form in which copper was traded, was also found.

The archaeometallurgical studies carried out on metal finds from Oman have confirmed Peake's intuition that the high content of nickel was the main feature that distinguished Omani copper ores. Copper ores rich in nickel, cobalt and arsenic were rather typical of copper deposits of the Semail Ophiolite, which characterize the Omani Mountains (Weeks 2003: 99-111). Metal produced from these deposits frequently contains that element as natural impurities (Hauptmann 1995: 246-248). According to statistical analyses, metal items from Oman and Bahrain regularly contain up to 4% of nickel and arsenic.

Because Bahrain has no copper ore deposits, it is possible to assume that all the metal objects that have this chemical composition were produced with copper from Oman (Prange *et al.* 1999: 191). Lead isotope studies on copper ores from Oman and on artefacts from Oman and Bahrain demonstrate that these archaeological items are fully compatible with Omani ores.

The same data were compared with the those from 180 metal ware from Mesopotamia dating from the Uruk, from the 4th millennium BC, to the Akkadian period, ca. 2300 BC, integrated with chemical compositions (Begemann and Schmitt-Strecker 2009). The result was that metal items with the signature of Omani copper occurs in Mesopotamia during all periods, but especially during the Early Dynastic III and the Akkadian period, when about half of the copper in use seems to come from Oman. It is intriguing to observe the absence of copper objects in the Gulf with a lead isotope signature linked to copper from India, even if written sources frequently associate all together Magan, Meluhha and Dilmun. It is possible that Dilmun was only a station on the trade route between Mesopotamia and India, and that the local market was reserved to the Arabian copper (Begemann *et al.* 2010).

Chapter 5

The earliest appearance of metalworking

The earliest copper objects started to appear in Omani settlements and graves around the beginning of the 4th millennium BC. During the 4th millennium BC, Southeastern Arabia – corresponding to the present-day Sultanate of Oman and United Arab Emirates – witnessed a substantial increase in the population density of the local groups of gatherers, fishermen and nomadic or semi-sedentary farmers. These communities progressively occupied the most productive areas of the region, creating highly specialized environments. Relying on the intensive exploitation of marine resources, the harbors set along the coast became soon centers of trade and craft activities. Oases developed inland with the creation of special artificial environmental systems based on the development of irrigation technologies, agriculture, livestock and trade with the coastal areas.

From the end of the 4th millennium BC rapid, intense social changes occurred in Southeastern Arabia. Considerable transformations took place not only in the economy, but also in the size, structure, and distribution of populations, at the outcome of Middle Holocene adaptive strategies. According to the archaeological evidence, trade networks were expanded and intensified considerably, mainly with Mesopotamia and Iran, in close connection with the early exploitation of the local metal ore resources located in the Omani mountains. However, the accretion of social complexity of the Southeastern Arabian people did not develop a 'state' apparatus with a centralized management, nor urban centers. Instead, it created a complex architecture of tribal alliances, despite what we observe in other countries of the Near and Middle East, where the formation of an urban civilization is a frequent result of the relevant socio-economic changes connected with a strong increase in productive activities. This peculiar socio-cultural evolution is testified by hundreds of monumental collective burials and by complex funerary practices (Bortolini and Tosi 2010).

During the 4th millennium BC Mesopotamia, Syria, Iran, Egypt and the Indus Valley had a revolutionary and drastic cultural and political transformation ending with the establishment of early states. These hierarchical, social structures organized human labor, the exploitation of resources and the accumulation of wealth. The new political organization of the Near Eastern countries acted as a catalyst for the Arabian communities, stimulating the transformation of their traditional subsistence activities into a more efficient exchange economy, in order to supply the international circuits with their most precious products, first of all copper (Cleuziou and Tosi 2007: 63-66). Anyway, they never modified their fundamental tribal political system, based on familiar bonds.

The first copper of Oman

Copper was a powerful stimulus for the development of Southeastern Arabia. It was in fact actively requested by the Mesopotamian economy that needed metals for its rapid development. Copper reached Mesopotamian cities from many sources, but the Arabian trade circuit could supply tons of metal from a relatively close source.

The earliest copper items from Oman were recovered in the physical remains of fishermen communities at coastal sites (Cleuziou and Tosi 2007: 90-91). Excavations at Ras Al-Hamra, in the most recent levels of RH-10, recovered some small copper-based items. Amongst them, the oldest copper object found in Oman. Moreover, many small copper fragments were found at the coastal site of GAS-1, a site at the mouth of Wadi Shab about 120 km south of Ras Al-Hamra. All these items were the remains of objects that had been worked with a non-specialized hammering technology. Other early copper fragments are known in the Ja'alan region from the sites of Ras Al-Jinz RJ-1 (Period I) and As-Suwayh SWY-2. The latter is a 3400 BC coastal settlement in eastern Ja'alan, south of Ras Al-Jinz, located on a sandbar immediately north of the channel that linked an old lagoon with mangrove swamps to the sea.

Most of the earliest copper objects were not ornaments or elements of personal decoration, but tools used in everyday life. Frequently they coexisted with typologically similar artefacts made of earlier materials, as stone or shell. Therefore, the new technology was starting to become strictly imbedded within the subsistence activities of the communities, making them partially dependent on metal trades that brought the copper to the coast from the production areas that were located far away in the mountains.

Wadi Shab GAS-1

Wadi Shab GAS-1 is a site located on the coast between Muscat and Sur a few kilometers north of Tiwi, at the mouth of Wadi Shab, on a cliff line facing the sea. Since the *wadi* connects the coast with the interior, this was a relevant area of human settlements not only for the presence of perennial fresh water, but also for the possibility of establishing contacts and trade. The local chlorite and steatite deposits were largely exploited for the production of stone ornaments. Radiocarbon dates suggest that the occupation of the site extended between the end of 5th and the mid of 3rd millennium BC (C14 dates, 2 σ: 4250 – 2700 cal. BC, 3800 – 3500 cal. BC). Evidence of huts and prehistoric working activities were identified; GAS-1 was also used as a burial ground. Prehistoric tombs dating to the 4th millennium BC were also excavated at this site. Wadi Shab GAS-1 was reoccupied much later, in the Islamic period (Usai 2006: 275-277).

The surface materials included a mixture of Late Islamic and prehistoric finds. A systematic, intensive collection was carried out on an area of 25 m², using a 1 x 1 meter grid (Tosi and Usai 2003: 8). The archaeological material recovered during the excavation were similar to the prehistoric objects recovered on the surface: lithic industry characterized by flake debitage production, limestone net-sinkers, soapstone earrings, finished and unfinished steatite beads, shell fishhooks. Together with this material, many small fragments of copper-based objects were recovered, most of them in the upper stratigraphic units (Tosi, Usai 2003: 11-20). All the metallic finds were attributed to a later time compared to all other materials (i.e. 3rd millennium BC), because of the conventional preconception that "bronzes" should necessarily be late. However, the archaeometallurgical studies carried out on the metal finds from GAS-1 showed that they are rather similar in chemical composition with the items from Ras Al-Hamra RH-10.

The collection of metals from Wadi Shab does not differ significantly from the same materials recovered at RH-10. Furthermore, a recent study carried out by M. Tosi and P. Koch on the surface finds from GAS-1, revealed that the archaeological indicators of craft activities – flint perforators, copper chisels and awls, unfinished soapstone beads, steatite fragments – were associated with the remains of stone huts.

These indicators are abundant in all the area, but they are scarce or even absent on the surface corresponding to the tombs: it means that the working layer containing the copper tools is earlier than the tombs, and therefore it has to be dated to the 4th millennium BC (Tosi, personal communication).

57 artefacts from GAS-1 were analyzed by XRF. All but two iron fragments were made with copper alloy; the corroded iron pieces were attributable to the later Islamic presence at the site. Copper items were made of almost pure copper, with an arsenic content that usually was lower than 1%, an indication that most of the objects were made of very pure copper as expected in such old specimens. Some of the objects with higher As content also had high Ni percentages: this phenomenon recurs frequently in Omani prehistoric metallurgy.

The analyses revealed that several items had high silver contents, ranging from 1 up to 8.6%. A similar composition occurred in the RH-10 copper finds, and it is probably an indication of a common provenance of this early copper. Only one single fragment, of the analyzed items, contained some tin: a fishhook that was produced in an uncommon copper-nickel-arsenic-tin alloy, characterized by low Sn and As contents, around 1% both, and high Ni content, around 7%. Because the hook was found on the surface, it could be later than the previous objects, even if no materials belonging to later prehistoric phases were collected at the site. Such low tin content does not effectively influence metal properties. According to some scholars, the limit for the distinction between unalloyed copper and tin bronze can be set at 1% (Pernicka *et al.* 1990: 272). Tin generally does not occur in copper ores, but there are some minerals that contain both copper and tin, as stannite, a sulfide of copper, iron, and tin. In Saudi Arabia, geological hydrothermal processes were responsible for the formation of stannite, chalcopyrite and arsenopyrite deposits (Kinnaird and Bowden 1991: 435). If the tin in the hook is an impurity, we should admit the existence of additional sources for the copper of Wadi Shab.

Types of copper artefacts from Wadi Shab GAS-1

Only a few types of metal artefacts were recovered at GAS-1, as we can expect from a very early metallurgical site (Figures 5.1 and 5.2). They were mostly tools, used as cutters and perforators, such as awls, chisels and drill points. There are also some objects devoted to fishing activities, like hooks and small netting tools for making and repairing fishing nets. At GAS-1 fishing hooks were also made with seashell. Very simple metalworking was practiced at the site, as indicated by the recovered fragments of unfinished copper items. The production was made only by hammering, to shape the imported copper pieces in order to obtain the objects they needed for their daily work. This basic technology possibly derived from Neolithic techniques used to obtain flint items.

Awls. The awls are the most common tool found at GAS-1: about half the examined metal items from the site. This is particularly intriguing because at Wadi Shab a large amount of steatite and chlorite jewelry was produced, mostly perforated beads. Awls are small bars with a thin, square section; they were made with almost pure copper (Cu 92-99%). Only three items had a low arsenic content, around 1%, probably unintentional and connected with the original ore; one reached a higher As content, 3%.
[Examined items: DA 32400_2; DA 32400_4; US101_002-003; US101_011-013; US101_ 020; US100_ 022-024; US100_ 027-029; US100_ 035-036; US100_ 0039-040; US100_ 0042-043; US100_ 046; US100_ 048; US 100_050].

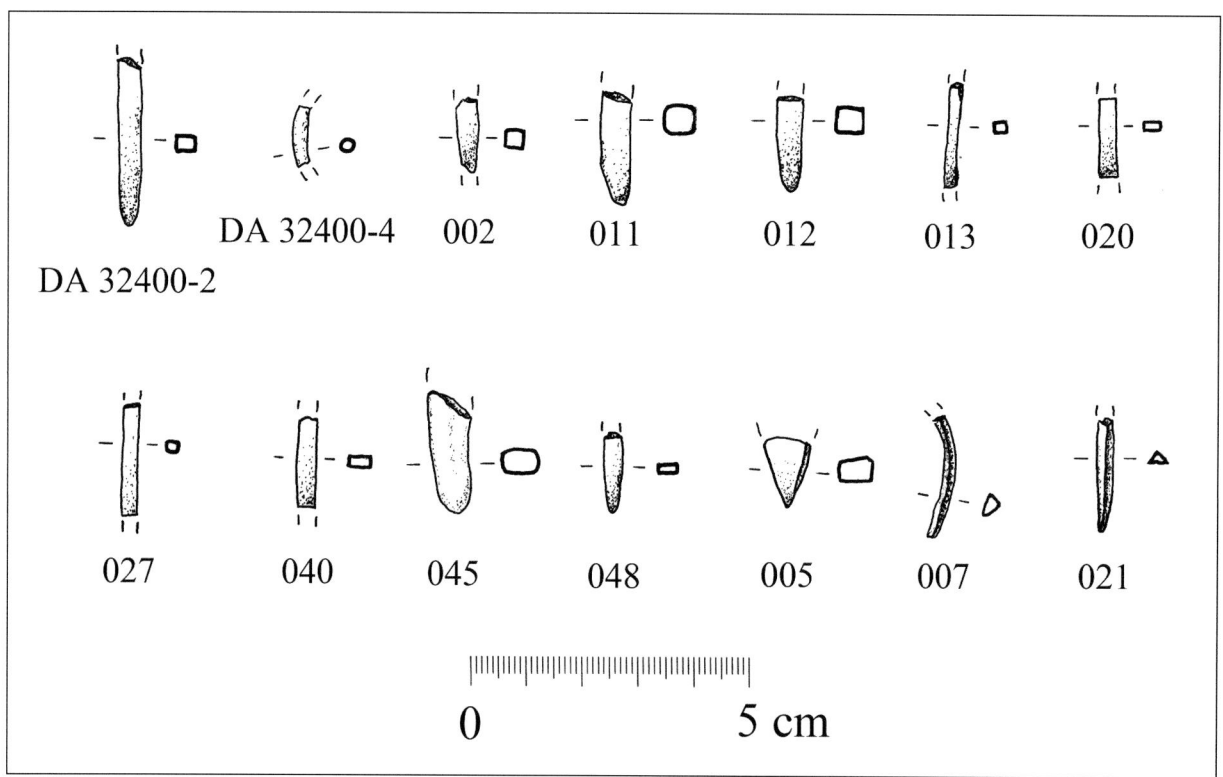

Figure 5.1. Metal objects from Wadi Shab GAS-1 (drawing by L. Tricarico).

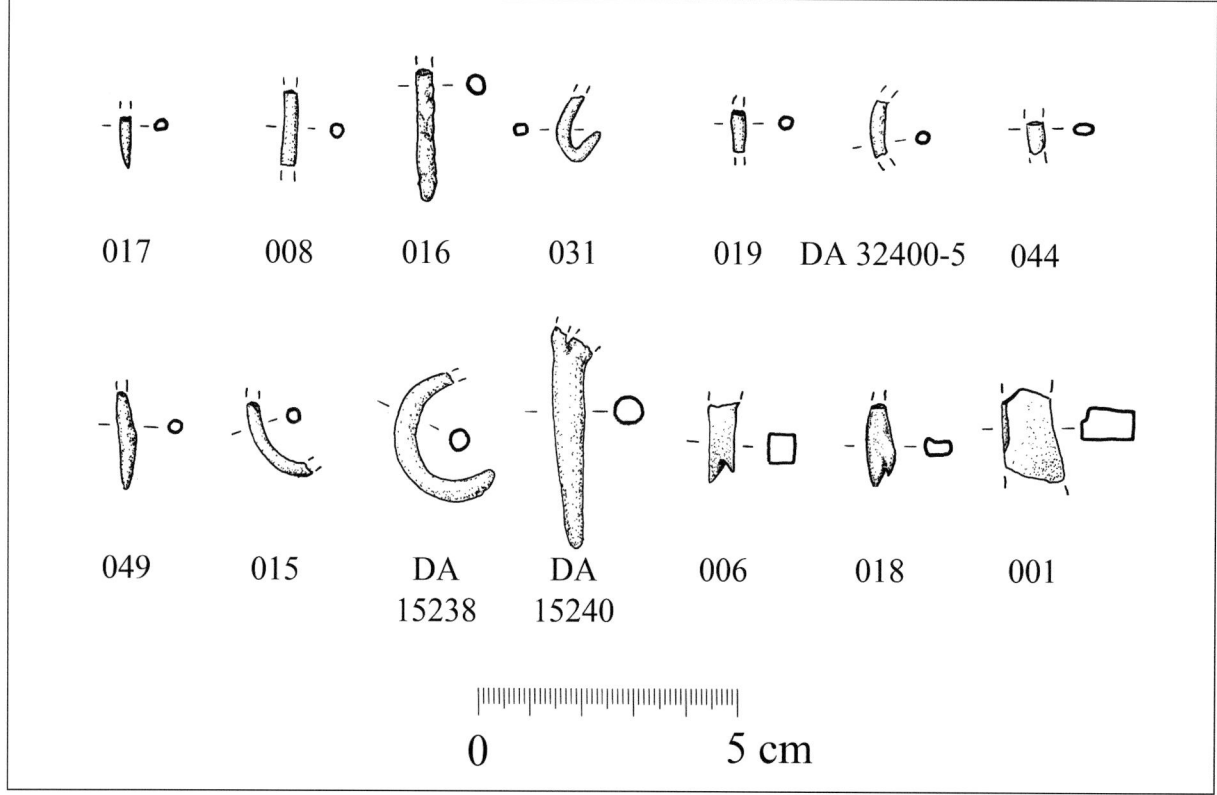

Figure 5.2. Metal objects from Wadi Shab GAS-1 (drawing by L. Tricarico).

Chisels. Chisels are thicker tools than awls, and therefore stronger and more resistant; their section is quadrangular and frequently rectangular. They were generally made of pure copper; only two items have an arsenic content exceeding 1%.
[Examined items: US100_ 004-005; US100_ 009; US101_ 014; US100_ 025-026; US100_ 033; US100_ 038; US100_ 0045; US100_0047].

Drill points. Thin points, triangular in section. They are rather small, about 1-2 cm in length, such as the double-backed perforators made of stone flakes that were the most common lithic tool found at the site (Tosi and Usai 2003: 11-12, fig. 7). One of the drill points is made in an unusual copper-silver-arsenic-nickel alloy (Ag 8.6%; As 3.3; Ni 2.9%); the other two in pure copper with silver impurities around 1%.
[Examined items: US100_ 007; US101_ 017; US100_ 021].

Pins. Small and thin pointed objects circular in section. They could be used as ornaments, or else as perforators with a round point. They were all made of unalloyed copper; one of them has a relevant silver content, around 5%.
[Examined items: US100_ 008; US101_ 019 (from a tomb); US100_ 051; DA 32400_3; US101_ 016; US 100_44; US 100_49].

Hooks. Hooks were made of almost pure copper, with the only exception of one item, DA 3240_5, which contained around 1% of tin and arsenic, 2% of silver, together with a large amount of nickel, 7%.
[Examined items: US101_ 015; US100_ 031; US100_ 041; DA 15238; DA 32400_5].

Netting tools. Little tools ending with two small spikes, probably used for making fishing nets. They were made of unalloyed copper.
[Examined items: DA 15240. Possible netting tools or unfinished items: US100_ 006; US101_ 018].

Unfinished items. These objects are an intriguing indication of a small-scale production of metal tools at the site, which used thin fragments of almost pure copper.
[Examined items: US100_ 032; US100_ 034; US100_ 037; US 101_001; DA 32400_1].

A prehistoric 'Swiss-knife' from Ras Al-Hamra

As it always happens, the emergence of metal use is characterized by rare and scarce evidence, also true in Oman. The few, small objects that were found in the early archaeological sites, mostly in a fragmentary state, testify that metal was still uncommon and somehow precious. One of the oldest metal artefacts from Oman seems to be a small 'knife' found at the late 4th millennium BC fishing encampment of Ras Al-Hamra RH-10, in the capital area of Oman. The site lies on a flat promontory that overlooks the mangrove shrubs at the mouth of Wadi Aday. The promontory, together with the Fahal Island – which lies only two miles in front of Ras Al-Hamra – forms a narrow sea channel for the passage of sardines that cross it in large shoals during winter. This special location offered control of the rich seafood resources to the fishing communities established there. The copper tool is 16.2 cm long and 4.4 cm wide (DA 26720), weighing 62 g and in appearance it looks like a small tanged knife, made by hammering (Figure 5.3).

Figure 5.3. Ras Al-Hamra RH-10. Multi-tasking tool (DA 2672) (photographs by P. Koch, courtesy Oman National Museum).

A careful examination of the wear traces preserved on its surface reveals that it was actually a multifunctional tool, produced to fulfill a variety of purposes and used for various tasks in daily life (Figure 5.4). Therefore, it was created according to the local tradition, still tied to the production of stone tools. The flat, double-edged blade shows traces of sharpening with a whetstone and it could be used for cutting, slicing and scraping.

Considering the context from which this object comes – a fishing village whose economy was based substantially on the exploitation of rich marine resources – the 'knife' use was mainly devoted to the world of fishing. The double-edged blade was used for both cutting and scaling large fish, and to sever lines, nets and ropes. Some indentations on the sharp end of the blade indicate that it was used to force some hard material too, maybe to remove shellfish, such as mussels, oysters or limpets from the rocks or to pry open seashells. The tool could be held in either way: the tang – a flat bar 5 cm long, rectangular in section – should not have been inserted inside a handle, but it had a specific function, because it is really a different tool (Figure 5.5). The end of the rod has in fact a light, but evident thinning step.

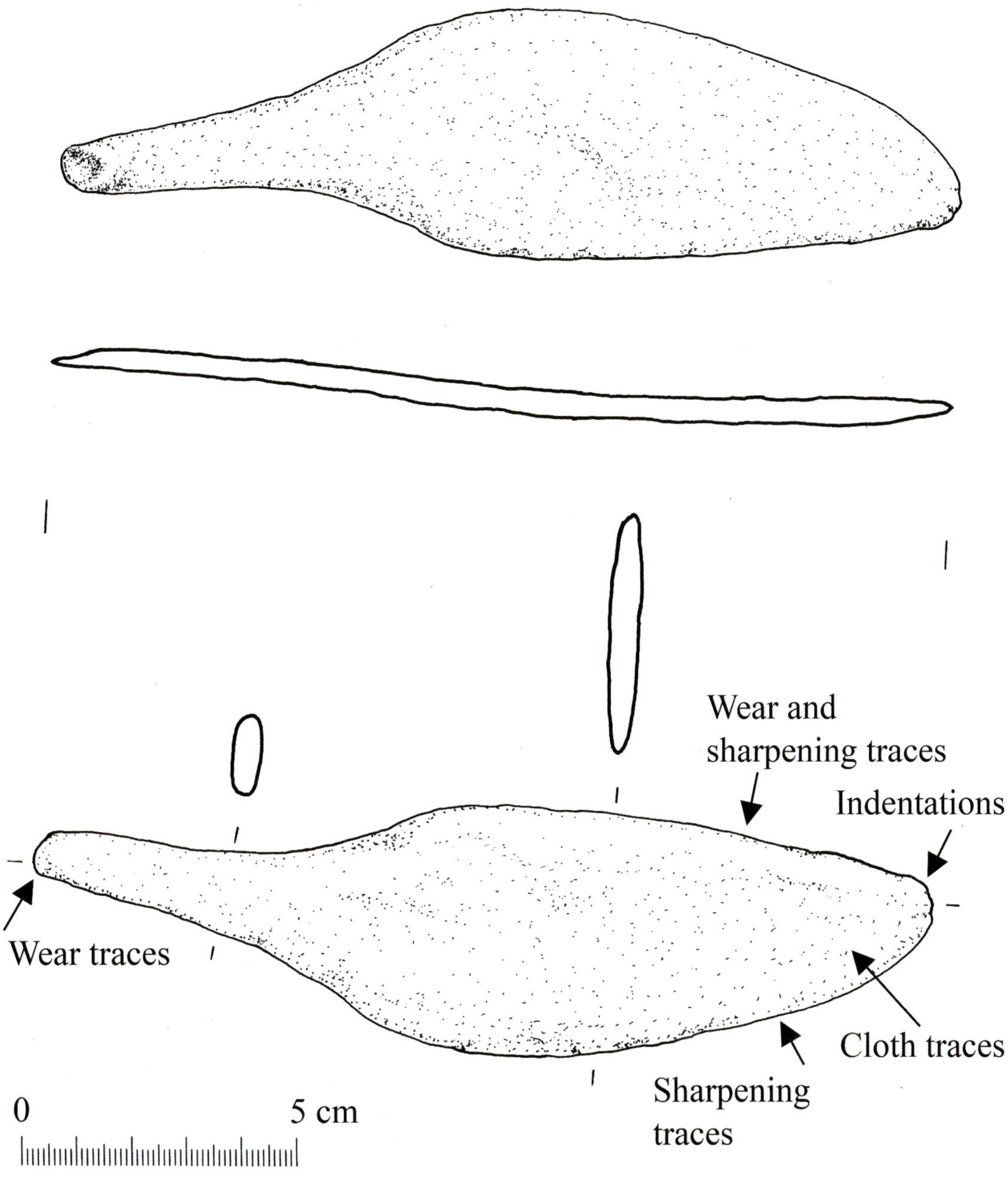

Figure 5.4. Ras Al-Hamra RH-10. Multi-tasking tool (DA 2672): arrows indicate wear and sharpening traces (drawing by L. Tricarico).

Figure 5.5. Ras Al-Hamra RH-10. Detail of the tang of the multi-tasking tool (DA 2672).

The presence of wear traces indicates that the tang was designed to be inserted between two hard surfaces, like a small, but strong wedge. It was probably used to open the valves of large mollusks by applying pressure in the delicate process of opening the shell, like a modern oyster-shucking knife. This kind of tool had to be functional and useful as a shell-opener: after centuries, similar copper tools were found in the Early Bronze Age fishing community of Ras Al-Hadd, at the HD-6 site.

The shape of these dedicated Hafit tools is morphologically similar to the 'tang' from RH-10 (for example DA 16493), but now the new instruments were devoted only to the function of shell shucking, because the multifunctional tools were substituted with specialized items. This multifunctional instrument – a kind of prehistoric 'Swiss-knife' – had a special value to the fishermen community, also because it was made of a very uncommon material as copper. Its high value is confirmed by the minute traces of mineralized textile fibers that were preserved in some spots, thanks to the thick green patina that covers the artefact. The fabric, which was in close contact with the metal, had to belong to a cloth container in which the object was kept.

Some other thin copper fragments were recovered at Ras Al-Hamra RH-10 (Figure 5.6). They were mostly small, simple tools for everyday activities in a fishing village: fishhooks, awls, chisels, drill points, pins. They were manufactured – or at least reshaped and sharpened – on site, as suggested by the hammer-stones and whetstones found during the excavation (DA 13310, DA 14326).

X-Ray Fluorescence analysis was carried out on 22 artefacts from RH-10. All items were produced with copper alloys; one corroded iron fragment was also found, belonging to a later presence in the area. Half of the objects were made of almost pure copper, with an arsenic content lower than 1%; the other half had a content higher than 1%, that reached values from 2% to 5% in three items only. It is hard and problematic to define when the presence of arsenic in a copper-based prehistoric object is only a natural impurity or an intentional addition in order to improve the characteristic of the metal, also because arsenic often occurs associated with copper ores. Relatively high arsenic contents connected with high percentages of other elements, as antimony, bismuth, nickel or silver suggests the use of a copper ore like *fahlertz*. Generally, many scholars suggest an arsenic content around 2 or 2.5% as the border line between unalloyed copper and arsenical copper (Giardino 2010: 184-185; Pernicka *et al.* 1990: 268).

Most of the objects with high As content also had high Ni content, an indication that these items were made of unintentional alloyed copper. Anyhow, there is at least one item (a small pointed tool) characterized by more than 5% of arsenic and a small amount of other impurities: it should be an early example of real arsenical copper. Many metal items from RH-10 have an unusual silver content, around 1-2%, which reaches a value of 6% in a chisel; most probably, these percentages are connected with copper ore.

Types of copper artefacts from RH-10

Awls. Awls are small tools with a square section. They were made of unalloyed copper; only one awl had an arsenic content as high as 3.5%.
[Examined items: DA 2639_06; DA 2639_11; DA 2639_15_1; DA 2639_19; DA 6595; DA 6973_35a; DA 6973_35b].

Chisels. Chisels are tools with quadrangular sections, stronger than the awls. The analyzed item had an unusually high silver content, around 6%, probably due to copper ore.
[Examined items: DA 10979].

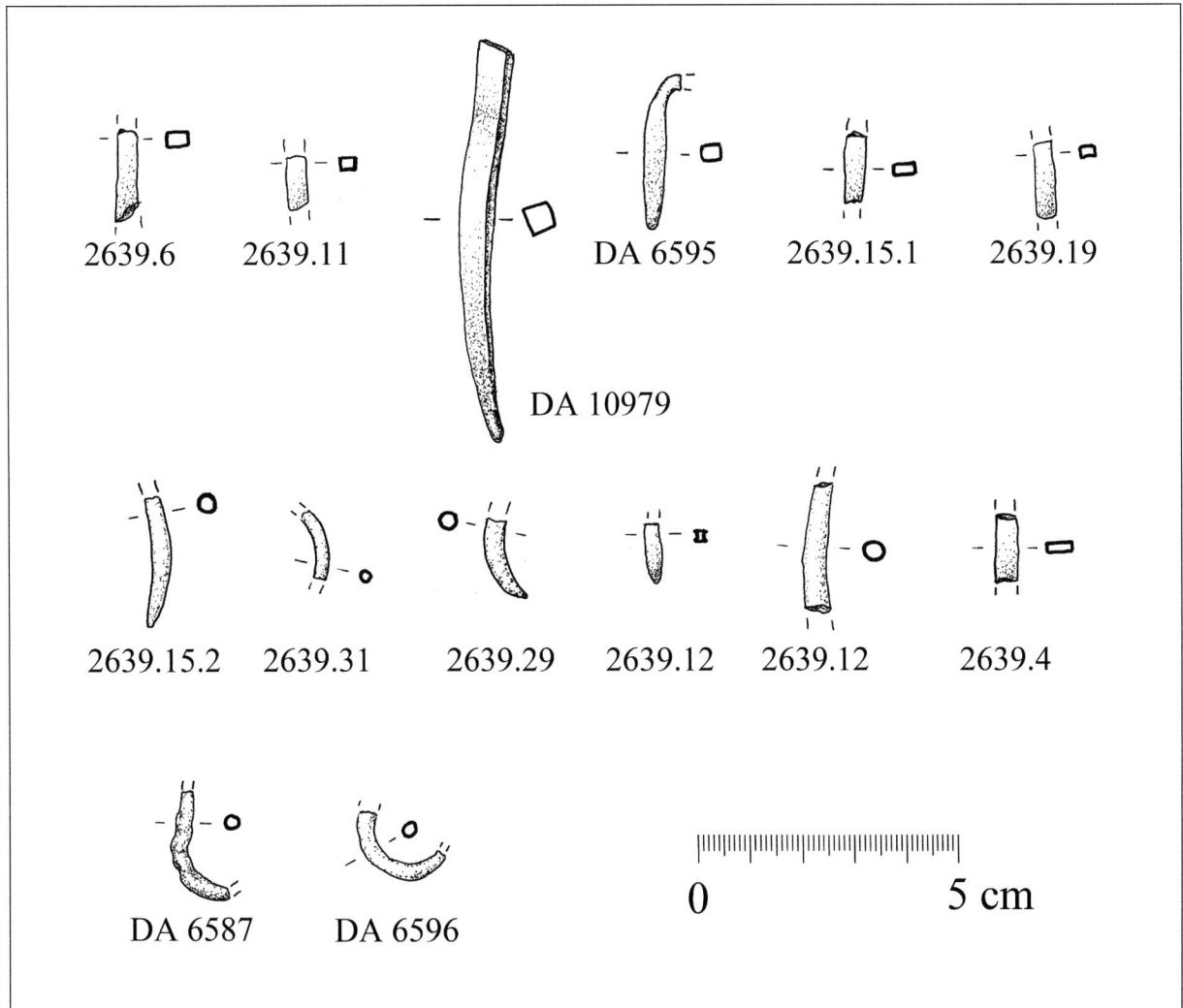

Figure 5.6. Metal objects from Ras Al-Hamra RH-10 (drawing by L. Tricarico).

Pins. Small objects circular in section. The analyzed pin was produced with unalloyed copper.
[Examined items: DA 2639_15_2].

Hooks. Hooks are a rather common kind of object in the metal finds from Ras Al-Hamra RH-10. They were produced with unalloyed copper.
[Examined items: DA 2639_09; DA 2639_29-31; DA 6587; DA 6596].

Pointed tools. Some tool fragments have this generic name because they are too small for a positive identification as awls, pins or even hooks. One of these tool fragments had a high arsenic content: 5.8%.
[Examined items: DA 2639_02; DA 2639_04; DA 2639_14; DA 2639_17-18, DA 2639_20].

Drill points. The examined drill point had a concave polygonal section, a peculiar shape obtained by hammering in order to improve the drilling capability. It contains a rather high silver content (around 2%).
[Examined items: DA 2639_12].

Problems and questions about the emergent metallurgy in Oman

Traditionally the advent of metals is considered as a great break from previous traditions. As in other regions, in Oman the evidence shows that at the beginning metal was used and worked as a 'stone', not only applying hammering techniques to shape the objects, but also creating multifunctional tools, as it was usual for lithic artefacts. Technological inventions and innovations – such as the use of copper instead of stone – must operate within the bounds of the world-view of the culture that receives them, if they want to be successfully accepted. A similar situation can be observed in Europe as well, where the first copper axes are close to the Neolithic polished stone ones, both in shape and in handling technique (Giardino 2010: 236). Unfortunately, we do not have analyses for these earliest copper implements from Oman, therefore, it is not possible to identify their provenance. There is no evidence of copper mining in the country at such an early period, therefore, it is reasonable to suppose that these first copper objects were made of imported metal, if they were not an import themselves.

Copper was in use since the 8th – 7th millennium BC in Mesopotamia and Anatolia, and copper smelting activities were identified in the Chalcolithic levels at Tal-i-Iblis (ca. 5500-3500 BC) and at Tepe Ghabristan in the Iranian Plateau (Pigott 1999: 76-77). Evidence for melting copper in a crucible was found at Mehrgarh, on the Kacchi Plain of Baluchistan, to the west of the Indus Valley, a pre-Harappan site dated to 4000-3500 BC. Large quantities of copper objects were recovered at Ganeshwar, a chalcolithic site in Rajasthan near Neem-Ka-Thana, in a context dating to the beginning of the 3rd millennium BC (Chakrabarti and Lahiri 1996: 32-36; Biswas 1996: 5-22; Kenoyer and Miller 1999). It is possible that the earliest metal came to the coastal sites of Oman from outside. There is clear evidence of contacts between Oman and Iran and Mesopotamia – both regions where metallurgy was already well developed – as early as the middle of 4th millennium BC.

Foreign clay vessels were found at Ras Al-Hamra, in the coastal settlement of RH-5 at Qurum near the present capital of Muscat, in a context of ca. 3400 BC; they were produced in Southeastern Iran and in Mesopotamia (Cleuziou and Tosi 2007: 87-88, fig. 64). Some copper could have been traded together with this pottery, starting the knowledge of this new material. Anyhow, only future analytical studies on the earliest copper items from Oman can provide scientific corroboration for any possible hypothesis. Nevertheless, it is also possible that copper extraction had already begun from the Omani copper ore veins, located in the mountain area.

Omani earliest copper finds are generally small and thin items: they show how much copper objects were uncommon and rare during that period. The sites of Wadi Shab GAS-1 and Ras Al-Hamra RH-10 are located along the southeastern coast, far away from the metal ore deposits; copper was not used to make precious ornaments, but to produce useful tools for everyday work. The presence of metal artefacts in the settlement is an indication that the local communities requested the copper; therefore, that metal had started to be part of the main exchange materials.

Another intriguing question is whether these early copper objects were produced with native copper or else smelted from ores by metallurgical processes. Unfortunately, the analytical identification of native copper in an ancient artefact is actually very uncertain and generally frustrating. A metallographic analysis may show the characteristic microstructure of native copper; nevertheless, the mechanical and / or thermal treatments – annealing after hammering, melting – generally changes the copper microstructure in the object, making difficult and uncertain the identification. In addition, early artefacts are often completely mineralized and therefore there is no metal for the characterization.

The study of the alloy chemical composition provides ambiguous answers too. The native copper is in fact very heterogeneous, although it is usually very pure. However, a remarkable purity characterizes also the copper smelted with primitive techniques from rich minerals, such as oxides and carbonates (Giardino 2010: 124-125).

Chapter 6

Early Bronze Age: the Hafit period, ca. 3200-2800 BC

A new era started in Oman at the very end of the 4th millennium, ca. 3200 BC, the Early Bronze Age. This period could be divided in two different phases, according to the funerary evidence, that distinguishes an earlier kind of burial, the Hafit type grave, from a later one, the Umm an-Nar type.

At the beginning of the Early Bronze Age, a large number of impressive collective stone burials, cairn graves, were built, the so-called called Hafit type graves (Figure 6.1). These monuments are spread all over the country and they were erected generally in highly visible places, in order to overlook the most strategic areas of the territory, to mark and to control the landscape. All the cairns are typically circular truncate-conical tower-like monuments, with a single chamber and a short corridor across the wall that faces East, towards the rising sun. They can be found as isolated tombs or grouped in necropolis. The Hafit type graves date between the last centuries of the 4th millennium BC and the first half of the 3rd millennium BC. About 3000 of them were identified just in the region of Ja'alan, sometimes grouped in cemeteries of hundreds of graves.

Figure 6.1. Hafit cairn burials from Jebel Misht, Wadi Ayn (photograph by C. Giardino).

The Early Bronze Age: the Hafit period, ca. 3200-2800 BC

Sometimes, a few metal objects were found inside the Hafit graves. They are small items, as copper rings, awls, needles and rivets suggesting that the graves were furnished with personal ornaments. At Al-Moyassar-25 (Grave 1), two copper needles were recovered together with a typical Hafit vessel, two flint flakes and several stone beads (Weisgerber 1981: 198-200; 2008: 1615, fig. 4). They are the earliest evidence of metal ware used as grave furniture in Oman and they are a clear indication that local communities promoted ornaments made of this material as relevant status symbols for the after-life. Another indirect clue of the relevance of metal in the Hafit society and economy is the presence, in the tombs, of foreign pottery from Mesopotamia. Sumerian pots collected from graves – such as the rim of a Jemdet Nasr type jar found at Ras Al-Jinz RJ-10, dated to ca. 3000 BC – are linked with the first establishment of copper trade from Magan to Mesopotamia recorded by cuneiform texts: they support and testify the prestige connected with foreign exchange (Cleuziou and Tosi 2007: 113-115).

The oldest remains of an agricultural settlement at Bat can be radiocarbon dated as early as 2800-2700 BC, a relevant evidence of the agricultural revolution that occurred in Oman at that time. With the creation of the oases system, farmlands were developed in areas where local climatic conditions never allowed farming without irrigation. Oases were formed from natural springs, underground creeks or aquifers, where water could reach the surface naturally by pressure or by man-made underground channels. Agricultural settlements of that type were established in the interiors at Bat, Amlah, Bisyah, Al-Moyassar, Buraimi, Ibri, Bahla, Nizwa, but also near sea lagoons, at the mouth of large *wadis* like Wadi Fulayj at Sur or Wadi Mijlas at Quriyat, mixing the resources of land and sea (Cleuziou and Tosi 2007: 139-141). The Bat oasis also has special relevance for the history of metal in Oman, because there is evidence of early metallurgy in this area.

The fishing village of Ras Al-Hadd HD-6 is probably the most meaningful settlement of the Omani Hafit period, not only from a metallurgical point of view, but also for its historical and archaeological aspects. The site dates from the end of 4th and to the beginnings of 3rd millennium BC: therefore, it is slightly later than RH-10. Ras Al-Hadd is a cape south of the city of Sur. HD-6 was located above a low sand dune, between the sea and an old shallow lagoon, which no longer exists, probably used as a harbor. It is bordered to the west and south by a steep cliff ca. 20 m high. The settlement had an irregular plan and a large wall, made of stones and mud-bricks, enclosed it. Inside the wall, the site was densely settled with mud-brick rectangular houses; they had from three to six small rooms, a few square meters each. Inside, remains of partially worked shells and soapstone testify that the buildings were also used to conduct some kind of household industrial activity (Figures 6.2 and 6.3).

The economy of the settlement was based on the exploitation of sea resources, particularly abundant in the coastal area of the Ja'alan region, thanks to the oceanographic phenomenon known as upwelling. Upwelling affects only restricted areas of the oceans, as in the northwest Arabian Sea. Each summer the strong south-west monsoon winds that blow parallel to the coast move the cooler, usually nutrient-rich deep water towards the ocean surface, replacing the warmer, usually nutrient-depleted surface water. This nutrient-rich water stimulates the growth of plankton that attracts a huge number of fish near the coast, increasing enormously the surface fishing productivity during Winter. Therefore, the real period of activity for the sailors of the region (Anderson and Prell 1993; Rai and Das 2011).

A huge amount of fish easily available in the sea near the coast stimulated the economy of the settlements located around Ras Al-Hadd, like HD-6. The sea products – not only fish, but also shells, dolphins and sea-turtles – were much more than the village needs for its subsistence, therefore they alimented an intensive trade with other areas of the country, bringing large amount of copper to the settlement: it is still easy to collect prehistoric copper fragments on the sand near the site.

Figure 6.2. Ras Al-Hadd HD-6. Three-dimensional reconstruction (image by M. Cattani, courtesy Joint Hadd Project).

Figure 6.3. Ras Al-Hadd HD-6. Building 1 under excavation (photograph by M. Cattani, courtesy Joint Hadd Project).

Many fish bones were recovered at HD-6, together with hundreds of hooks, many of them made with copper. The use of copper to produce hooks – a kind of tool that can easily get lost in catching fish – is an indication that this metal was now common and easily available, thanks to the large-scale fishing activities carried out at the site. The sea products were therefore first fished, then prepared and preserved in order to be traded with the villages in the interior of the country that exchanged them with other products, including copper.

Ras Al-Hadd HD-6: the earliest evidence of copper manufacturing

The metal objects from Ras Al-Hadd HD-6 are the best known archaeological metal complex from the Hafit period. The site provided the earliest evidence of *in situ* manufacturing of copper artefacts: in the settlement, many metallic artefacts at different production stages were recovered during the archaeological excavations. They have been the object of a multi-year archaeometallurgical study. The items were first classified and many were carefully analyzed, using a multi-disciplinary methodology of investigation, mainly based on non-destructive and non-invasive analysis techniques. Hundreds of metal items were collected at this site; they were made of pure copper or arsenic-copper alloys, sometimes with the intentional addition of nickel. In the village, the metal was in use for making weapons like daggers, but mostly for tools: hooks, pins, awls and chisels for shell and stone working, points to open shells and micro-forks for fishing-nets repairs.

There is no evidence at all of copper melting from the Hafit sites excavated along the Omani coast: metal was worked only with mechanical techniques, cutting the imported ingots in small blocklets and then shaping them by hammering. The large amount of ingot fragments and unfinished tools testify to the importance of HD-6 as an early metal working site. The ingot fragments recovered are also a significant indication of a large import of copper from smelting sites, in the inland part of the country. This is a relevant clue that a strong exchange network was already established between the coastal sites and the oasis settlements situated near the metalliferous outcrops, most probably thanks to a complex system of tribal social relations.

Many HD-6 metal artefacts were analyzed, using different techniques. Archaeometallurgical analyses were mainly carried out by X-Ray Fluorescence and Scanning Electron Microscope; unfortunately, some of the artefacts were poorly preserved, with little or no metal left. A first group of archaeological items from HD-6 was analyzed in Rome during 2006, using a portable X-Ray Fluorescence system.[1] The results are reported in Tables 6.1 and 6.2: data are quantitative and semi-quantitative (Giardino *et al.* 2007) (Tables 6.1 and 6.2). More recently, in 2012 and 2013, another group of metal finds was analyzed in Naples, with a more developed X-Ray Fluorescence system, and the results are presented and discussed in Chapter 12.

---

[1] A Spectro Midex X-Ray Fluorescence system was used to analyze the complex of the metalwork. It had the following technical characteristics: tube: Mo anode; Hv max 45 kV; power anode, max 0.35 mA; collimator diameter, 0.7 mm. The characteristics of the detector are: SDD (Silicon Drift Detector) cooled by a Peltier cell; resolution from 150 eV to 6.4 keV; a multi-channel: 2048 channels. Pointing system: two laser diodes. Dead time < 20%. The measurements lasted 100 seconds each. The system was self-calibrated and the quantitative analysis was fulfilled with a proprietary program using the Fundamental Parameter Approach. The system was kept under control with a set of reference materials. The standard deviation of the results obtained were within the limits reported. The standard deviation of the results was: 10% for concentration from 0.1% to 3%, 5% for concentrations from 3.1% to 20% and 1% and less for concentrations from 20.1% to 100%. The result of the XRF analyses were preliminary published in the Proceedings of the 2nd International Conference "Archaeometallurgy in Europe 2007". Then the analyses were revised for nickel content. This new version of the data was presented at Ravenna during the 19th Conference of the European Association of South Asian Archaeologists on July 2007 as a poster titled "Archaeometric Data from the prehistoric Site of Ras Al-Hadd HD-6 (Sultanate of Oman)", by C. Giardino, G. Guida and S. Ridolfi.

DA	Description	Zn	As	Pb	Sn	Fe	Ag	Ni	Cu
15684/02	Awl	0,0	1,2	0,0	0,0	0,0	0,2	3,5	95,1
14984	Blocklet	0,0	0,1	0,0	0,0	0,2	0,0	0,0	99,6
15071	Blocklet	0,0	0,2	0,0	0,0	0,1	0,0	5,3	94,5
16465	Blocklet	0,5	0,6	0,0	0,0	0,0	0,1	0,7	98,1
16467	Blocklet	0,0	0,9	5,8	0,0	0,0	0,1	0,0	93,2
16472	Blocklet	0,0	0,3	0,0	0,0	0,0	0,2	0,0	99,5
15683/01	Chisel	0,0	4,3	1,1	0,0	0,0	0,2	3,0	91,4
15663	Dagger	0,0	4,7	0,0	0,0	0,0	0,0	4,1	91,3
15075	Pin	0,0	4,2	0,0	0,0	0,0	0,1	4,0	91,8
16464	Pin	0,0	7,5	0,0	0,0	0,0	0,3	0,0	92,2

Table 6.1. ED-XRF quantitative and semi-quantitative analyses of Ras Al-Hadd HD-6 metalworks.

According to these results, we can distinguish two main copper alloys in the HD-6 finds; copper content is almost unique in one of them, while in the other one copper is alloyed with arsenic. In both cases, nickel is a common, usual presence, which also frequently occurs in rather high contents. Some of the objects from HD-6 were made with arsenical copper, a special copper alloy whose use was peculiar to early metallurgical cultures. This is the oldest form of copper alloy, followed by tin bronze only much later: arsenical copper alloys spread during the initial stages of metallurgy, in a period between the late 4th and early 3rd millennium BC, even though they were still in use during the 3rd millennium, when the use of tin-bronze was being established.

The addition of arsenic to copper lowers the melting point; it also acts as a deoxidizer, facilitates pouring the metal and improves its mechanical properties. It increases the breaking load considerably, and allows cold forging. It neutralizes the effect of certain damaging impurities too, like bismuth. Smelting copper ores rich in arsenic or copper arsenate was the most likely way to get arsenical copper. Probably one of the main reasons that drove early metallurgists to prefer arsenical copper to pure copper was the different mechanical behavior when the alloy was submitted to cold working. However, both pure copper and arsenical copper have similar characteristics as cast state, after cold hammering the copper arsenic becomes much harder and more resistant than the pure copper: it is in fact comparable to bronze. It also makes easier both cold and hot working. Besides, we must not underestimate the aesthetic effect: rich arsenic alloy has a nice silvery appearance; the surface of the object is enriched of arsenic because of segregation, it has therefore a pleasant pale hue without weakening the artefact. It is hard and debatable to establish if the addition of arsenic in the alloy was intentional: it is possible that low percentages of arsenic (less than 2.0-2.5% As) are related to the ore, and therefore unintentional; an alloy with relevant arsenic contents – with average (2.5-6.0% As) and high values (greater than 6.0% As) – seems to show that the ancient metallurgist wanted to get specific characteristics for the objects (Giardino 2010: 184-185).

Some of the finds from HD-6 have an arsenic content around 4-5%: significantly, they were chisels, awls and daggers, all objects that would need a stronger material. There are also items where some arsenic ore was added to the melted copper to obtain the alloy. Most probably, some peculiar arsenic copper ore was deliberately selected to produce a natural alloy that the metallurgist used to make specific objects: the metal color was probably of fundamental help. The Omani copper ores are significantly rich in arsenic and nickel and the average concentration of both elements is comparable in ore and metal objects not only to Hafit items, but also to later finds at Umm an-Nar and Wadi Suq materials (Prange 2001: 95).

The Early Bronze Age: the Hafit period, ca. 3200-2800 BC

DA	Description	Zn	As	Pb	Sn	Fe	Ag	Ni	Cu
15084	Awl	-	Tr	-	-	-	-	++	MC
11545	Awl	-	-	-	-	-	-	+	MC
12640	Awl	-	Tr	-	-	Tr	Tr	++	MC
15051	Awl	-	+	-	-	-	Tr	+++	MC
16481	Big Chisel	+	Tr	-	-	Tr	-	0	MC
11533	Blocklet	-	++++	Tr	-	Tr	-	+++++	MC
11536	Blocklet	-	-	-	-	Tr	-	0	MC
11543	Blocklet	+	+	-	-	-	-	+++	MC
11554	Blocklet	-	-	-	-	-	-	++	MC
11590	Blocklet	-	Tr	-	-	-	-	++++	MC
11596	Blocklet	-	++++++	-	-	Tr	-	+++	MC
17191	Blocklet	-	-	-	-	Tr	-	++	MC
17189	Crochet	-	Tr	-	Tr	Tr	Tr	0	MC
11553	Awls	-	Tr	-	-	++	-	0	MC
17204	Crochet	-	-	-	-	-	-	+++++	MC
17205	Crochet	-	++++	Tr	-	Tr	-	0	MC
17224	Crochet	Tr	Tr	-	-	Tr	-	++++	MC
15664	Dagger	-	++++	-	-	-	-	++++++	MC
11541	Drill Point	-	Tr	-	-	Tr	Tr	++	MC
11569	Drill Point	-	Tr	-	-	Tr	-	0	MC
11588	Drill Point	Tr	++	-	-	Tr	-	++++	MC
11599	Drill Point	-	+	-	-	Tr	-	++++++	MC
17194	Drill Point	-	+	+	-	Tr	-	++++++	MC
17214	Drill Point	-	+	-	-	Tr	-	+++	MC
17230	Drill Point	-	Tr	-	-	Tr	Tr	0	MC
17233	Drill Point	-	Tr	-	-	+	-	++++++	MC
15683/02	Flat Chisel	Tr	+	-	-	Tr	-	++	MC
11546	Fragm. Awl	-	Tr	-	-	-	-	+++++	MC
15079	Fragm. Awl	-	+	-	-	-	-	0	MC
17190	Fragm. Awl	-	Tr	-	-	-	-	0	MC
11591	Fragment	-	Tr	-	-	Tr	Tr	++	MC
14120	Hook	-	++	-	-	-	-	++	MC
15660	Hook	-	Tr	-	-	-	-	++++++	MC
16443	Hook	Tr	+++	-	-	-	-	+++	MC
14987100	Hook	-	Tr	-	-	++	-	++	MC
15667/06	Hook	-	Tr	-	-	-	-	0	MC
15667/07	Hook	-	Tr	-	-	Tr	-	++	MC
15654	Pin	-	Tr	-	-	Tr	-	0	MC
11555	Pin	-	++++	Tr	-	Tr	-	+++++	MC
16510	Pin	-	Tr	-	-	Tr	Tr	+	MC
17192	Pin	-	Tr	-	-	Tr	-	++	MC
11573	Sheet frag.	-	Tr	-	-	+	-	+++	MC
16493	Shell-opener	+	-	-	-	-	-	0	MC
14707/03	Shell-opener	+	+	-	Tr	Tr	-	++	MC
11547	Small Awl	-	+	-	-	+	-	0	MC
11589	Small Awl	-	+	-	-	Tr	-	++	MC
14121	Unfin.Hook	-	+	+	-	Tr	-	++	MC
11535	Unfin. object	-	Tr	-	-	-	-	++	MC
17203	Unfin. object	-	+++++	-	-	Tr	-	0	MC

Table 6.2. ED-XRF quantitative and semi-quantitative analyses of Ras Al-Hadd HD-6 metal works. Semi-quantitative composition: MC = main component; ++++++ → + diminishing content; Tr = trace (modified after Giardino *et al.* 2007).

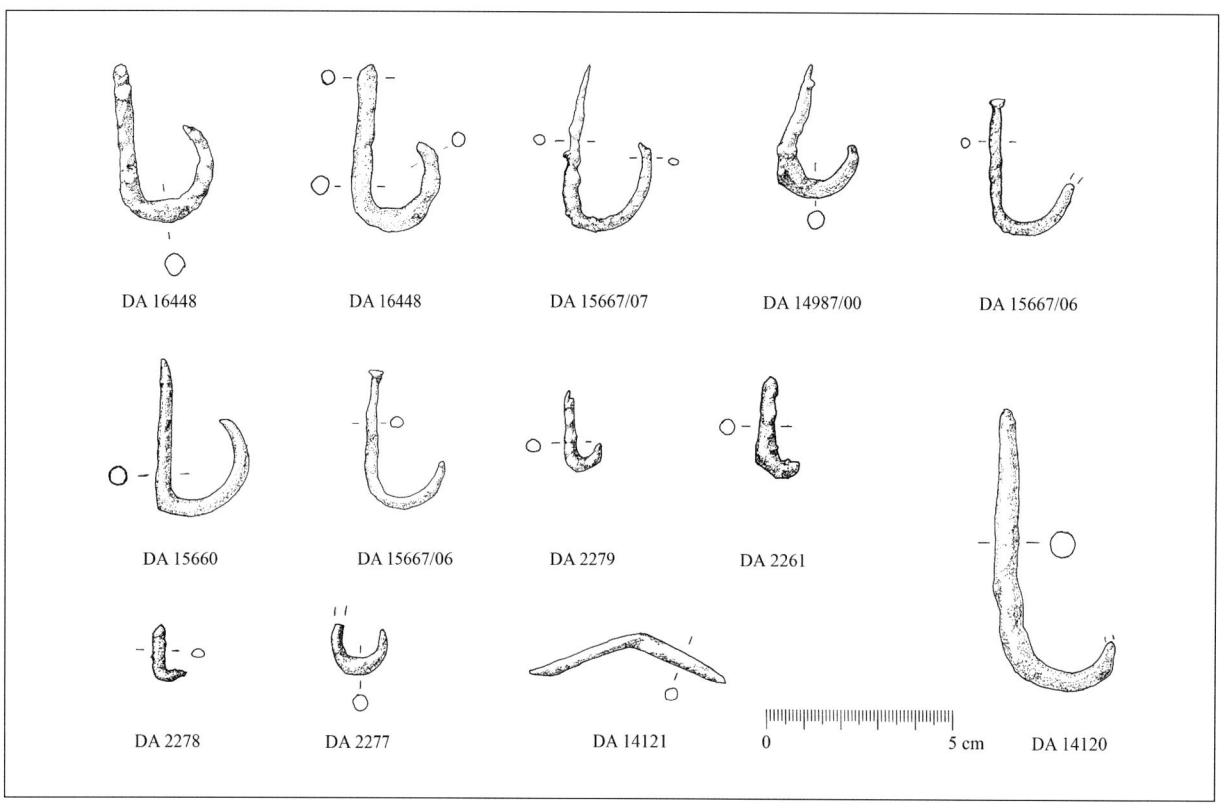

Figure 6.4. Ras Al-Hadd HD-6. Fishhooks (drawings by L. Tricarico).

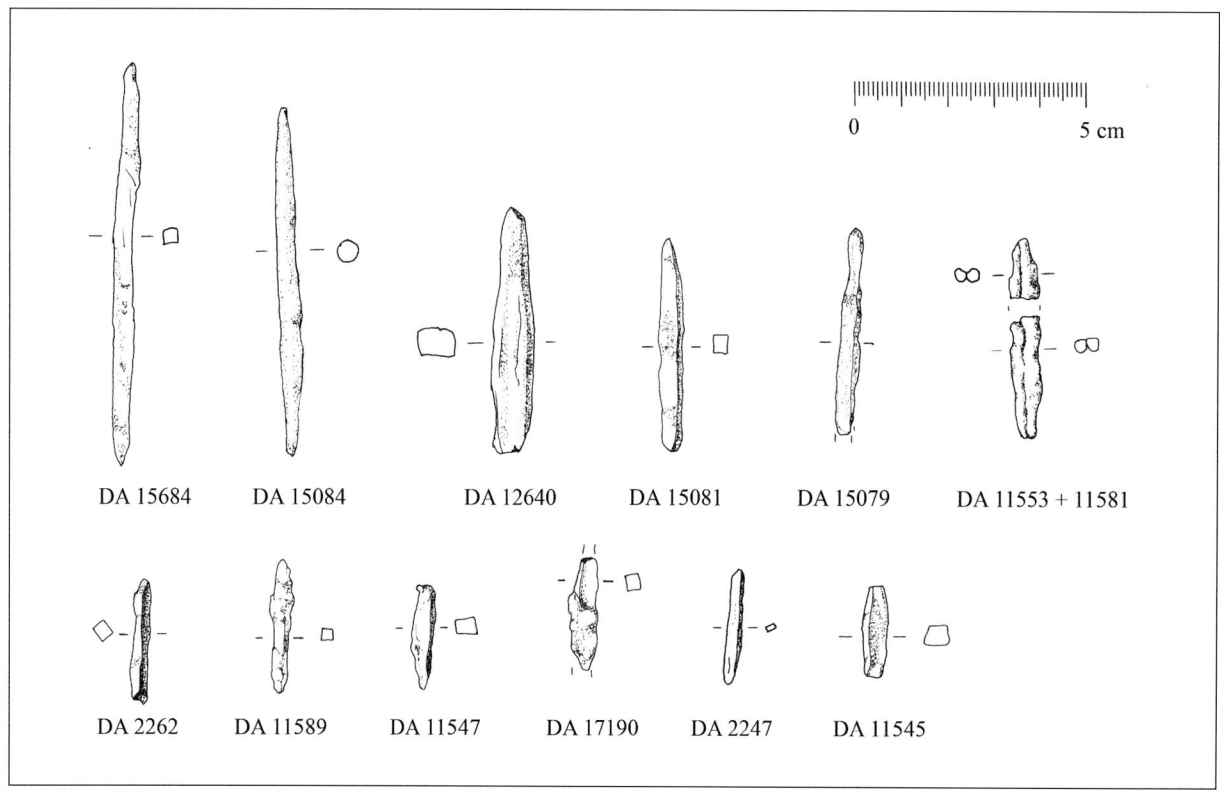

Figure 6.5. Ras Al-Hadd HD-6. Awls (drawings by L. Tricarico).

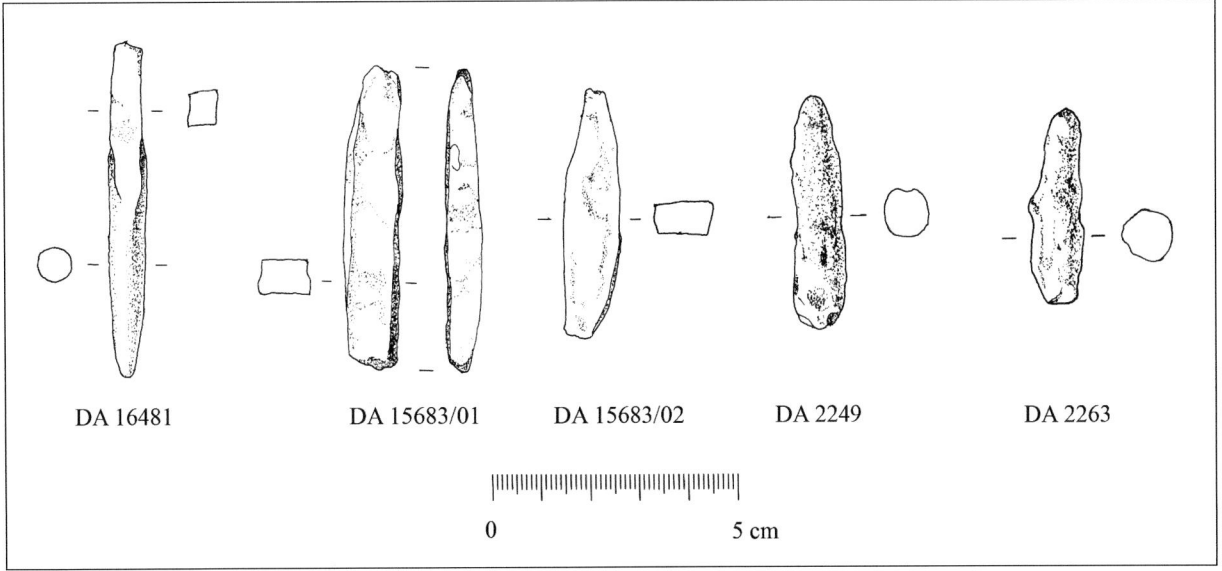

Figure 6.6. Ras Al-Hadd HD-6. Chisels/punches (drawings by L. Tricarico).

Types of copper artefacts from Ras Al-Hadd HD-6

The metal artefacts found at HD-6 have very little typological variation: the large majority were tools devoted to fishing activities. There is also a remarkable number of unfinished items that testify to the existence of workshops – at household level - that provided the production of the copper objects used in everyday life.[2] Only a few categories of objects were discovered in the hundreds of metal finds recovered from the site; the presence of very few types of ornaments and weapons stresses that the village activities were almost entirely devoted to the exploitation of sea resources.

Tools

Tools are the most common metal production at Ras Al-Hadd HD-6. According to elementary analyses carried out on these items, they were generally made with almost pure copper.

Fishhooks. In the Ras Al-Hadd settlements, where line fishing is attested since the Neolithic period, fishhooks were crafted not only in copper alloy, but also in other materials including seashell. Copper fishing hooks show some sort of typological diversity: they are generally similar in shape, but they differ quite a lot in their size, according to the kind of fish they were designed for (Figure 6.4). They testify the presence of very specialized fishing, because hooks were manufactured according to actual needs. Their length is very variable, ranging from 1.5 up to 11.5 cm; the average length is about 3-5 cm. The shanks are always straight, circular in section; their diameter (including corrosion) ranges from 3 up to 7 mm.

---

[2] This brief typology reflects the metal finds from the site. Because of the large number of copper-based objects found at HD-6 during the excavations, the following inventory numbers (DA) refer only to the items that were examined; they were selected as peculiar and characteristic. Most of them were analyzed by X-Ray Fluorescence.

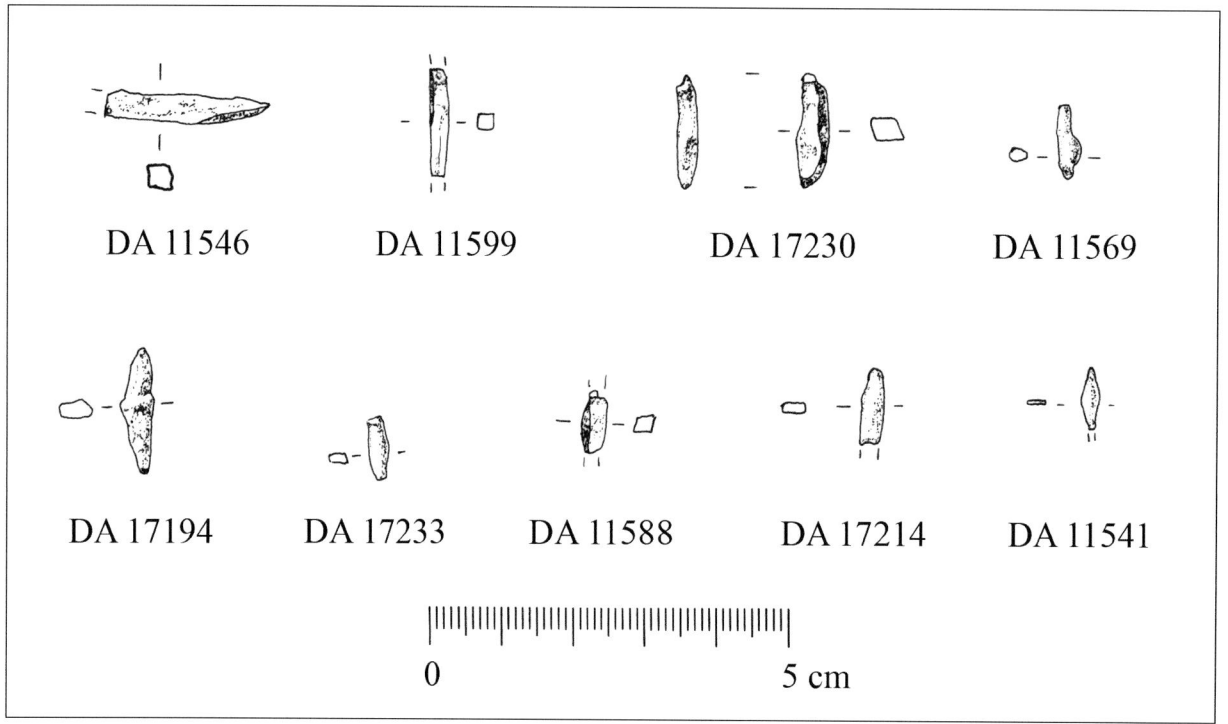

Figure 6.7. Ras Al-Hadd HD-6. Drill points (drawings by L. Tricarico).

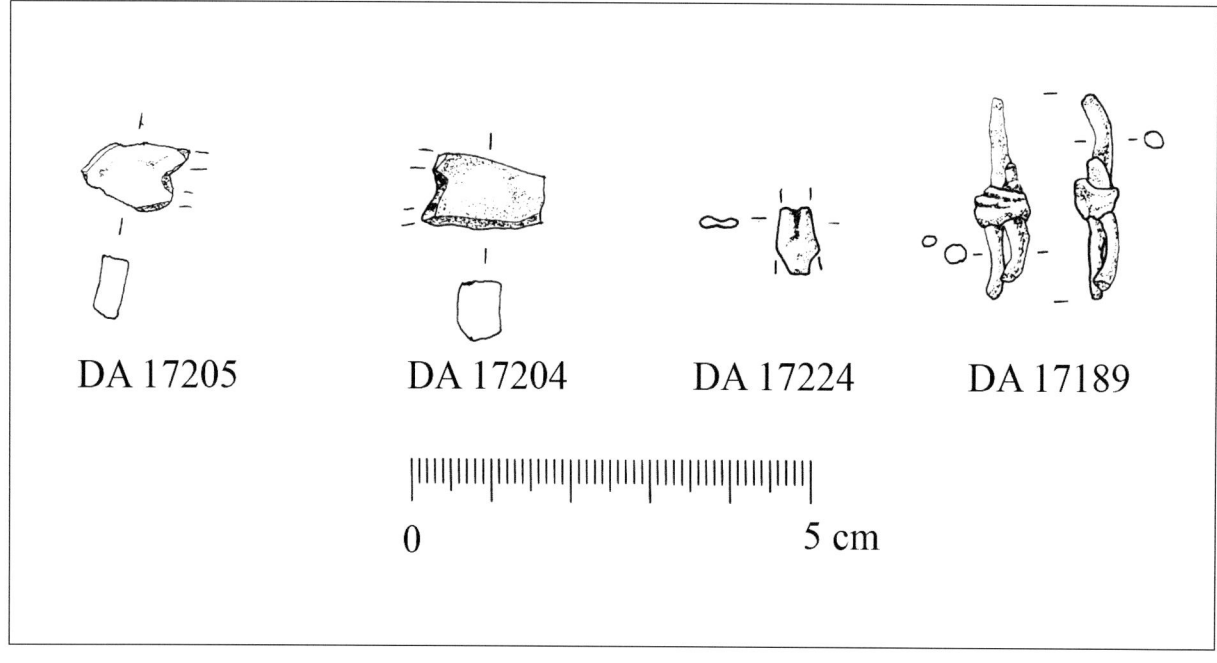

Figure 6.8. Ras Al-Hadd HD-6. Crochets/netting needles (drawings by L. Tricarico).

The less corroded items have a round, multifaceted section; they clearly show that they were produced by hammering starting from a square bar. In most of the fishhooks, the shank is connected to the point by a slight bend; but there are some items where the connection produces an elbow bend. Usually the straight shank ends with a point; only in few pieces, the end is flattened, in order to better fix the fishing line. Probably the manufacturing process to obtain fishhooks started from a blocklet; at a later stage, it was hammered in order to obtain a sort of angled awl. A possible semi-finished product is a sort of "boomerang tool", like DA 14121: a small bar 5.5 cm long, 0.4 cm wide, circular in section, slightly curved in the middle. According to compositional analyses hooks were mostly produced in almost pure copper; some have an As content up to 2%; median Ni concentration is 0.2%.
[Examined items: DA 2261; 2277; 2278; 2279; 11578; 12484; 14120; 14702; 14981; 14987/00; 15660; 15667.01; 15667.06; 15667.07; 15667.08; 16448. Uncertain attribution: 14121 (a sort of "boomerang" instrument, or, most probably, a semi-finished fishhook)].

Awls. The awls are rather common in the items of HD-6 (Figure 6.5). These tools had many different uses: for making holes in wood, for piercing holes in leather, for working materials such as soapstone or shells. Some of them are squared in section, while others are round. The awls were made from rods that were ca. 0.3-0.4 cm in section; with both ends down to fine points: the tool had therefore two pointed ends. The length is variable, ranging from ca. 4 cm up to 8-9 cm. Most awls were produced in almost pure copper. Only DA 14707.01 has a remarkable arsenic content, 2.4% As, associated with 3.25% Ni: it was made by an arsenic-nickel alloy. A small item (DA 2247) has an uncommon silver content, 2.4% Ag: it could be a pin fragment too.
[Examined items: DA 2262; 8631; 11545; 11547(small); 11553 + 11581 (couple of awls joined together); 12640; 14707.01; 14724; 15073; 15076; 15079 (fragment); 15081; 15084; 15684.02; 17270; 17190 (fragment). Uncertain attribution: DA 2247].

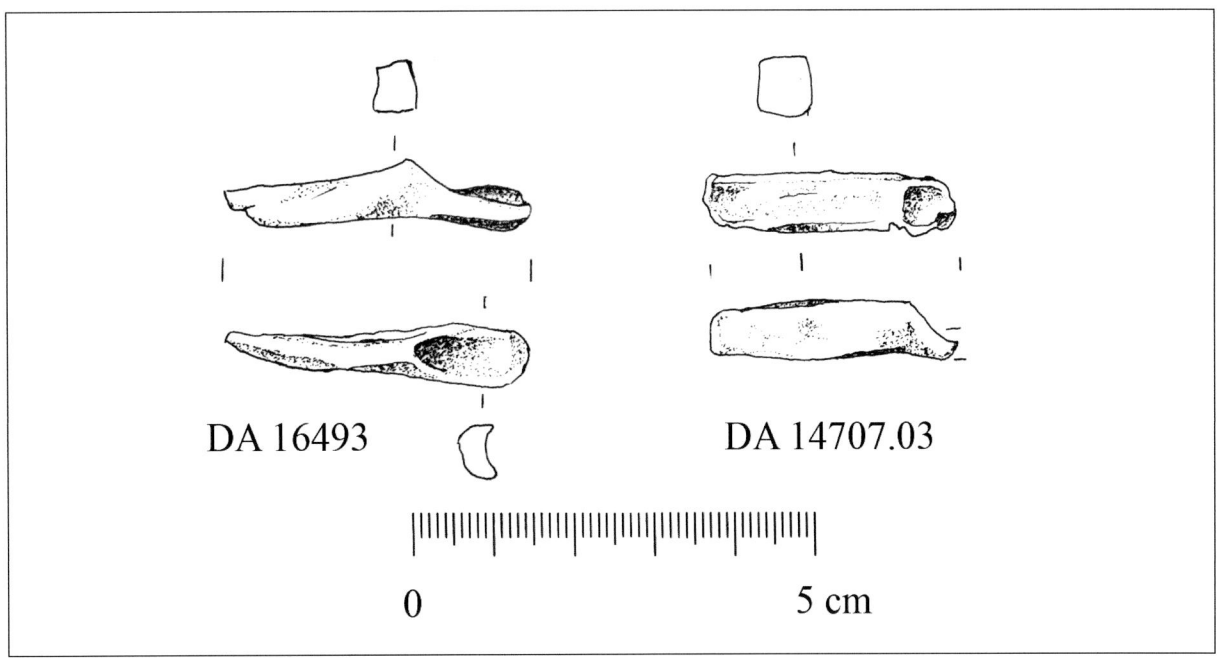

Figure 6.9. Ras Al-Hadd HD-6. Shell-opener (drawings by L. Tricarico).

Chisels/Punches. Chisels are rather similar to small awls, but much thicker, in order to obtain a more resistant tool (Figure 6.6). The end was in fact devoted to cut or pierce hard material such as strong wood, stone, or even metal by hand, struck with a stone mallet. They could be also used as punches. They are generally not too long: the average length is about 4-6 cm; their diameter is ca. 0.7-0.8 cm. Some are round in section, while some have a squared tang – to be inserted in a handle – with the end finishing in a circular point. The chisels analyzed were made of almost pure copper, around 99% Cu.
[Examined items: DA 2249; 2263 (small); 15072; 15078; 15683.01; 15683.02 (flat chisel); 16481 (big chisel); 17271 (flat chisel)].

Drill points. They are small and thin points, quadrangular in section, 1-2 cm long (Figure 6.7). They were held in a wooden drill, which rotates them, providing an axial force to create a hole on hard materials. From this period on, copper drill points started to substitute the stone "perforating implements" that occurred in Oman in the 4th millennium BC at sites such as at Ras Al-Hamra RH-5 (Usai 2005). Some of them were probably the point of awls and pins, whose reuse was determined by the need to exploit the metal as much as possible. Some others could come from sharp bit residues from ingot cutting.
[Examined items: DA 11541; 11546; 11569; 11588; 11599; 17194; 17214; 17230; 17233].

Crochets/Netting needles. This small, bifid tool is generally a fragment of a small rectangular bar ca. 1 cm long, 0.2-0.4 cm wide, cut in the middle at one end in order to produce a sort of swallow tail (Figure 6.8). The instrument is peculiar to HD-6 and most probably, it is connected with fishing activities. According to archaeological and ethnographic parallels of similar objects made of wood or bones (Magdalensberg and Gostenčnik 2010), it could be used, after handling, as a crochet to make and to repair fishnets, because its shape is very similar to modern netting needles. Probably another type of crochet is a tool made with a bundle of two copper wires; they were joined together with thin wire (DA 17189). Some of them show a relevant nickel content, about 1% Ni.
[Examined items: DA 15088; 15091; 15092; 17189 (bundle of wire); 17204; 17205; 17224].

Shell-openers. This very specialized tool is linked to the economy of the settlement, devoted to the exploitation of sea resources (Figure 6.9). It is made by a small rod ca. 3.5 cm long, flattened at the end in order to create a step. They could be used to pry between the valves of large mollusks like a modern oyster-shucking knife. They are the evolution of the multifunctional tool from RH-10 (DA 26720).
[Examined items: DA 14707.03; 16493].

Ornaments

Pins. Typically pins have a long shaft and a sharp tip; the sharpened body penetrates the material for which the pin is designated for (Figure 6.10). Pins are one of the few objects from HD-6 that can be regarded as an ornament, and the pins from HD-6 are probably the earliest objects used for decoration found in Oman. They look like awls, but they are rather longer: their length ranges from ca. 12 cm up to 14 cm. Pins are square in section, that is ca. 3-4 cm thick. They have two pointed ends: most probably, a perishable head was originally added at one end.

The Early Bronze Age: the Hafit period, ca. 3200-2800 BC

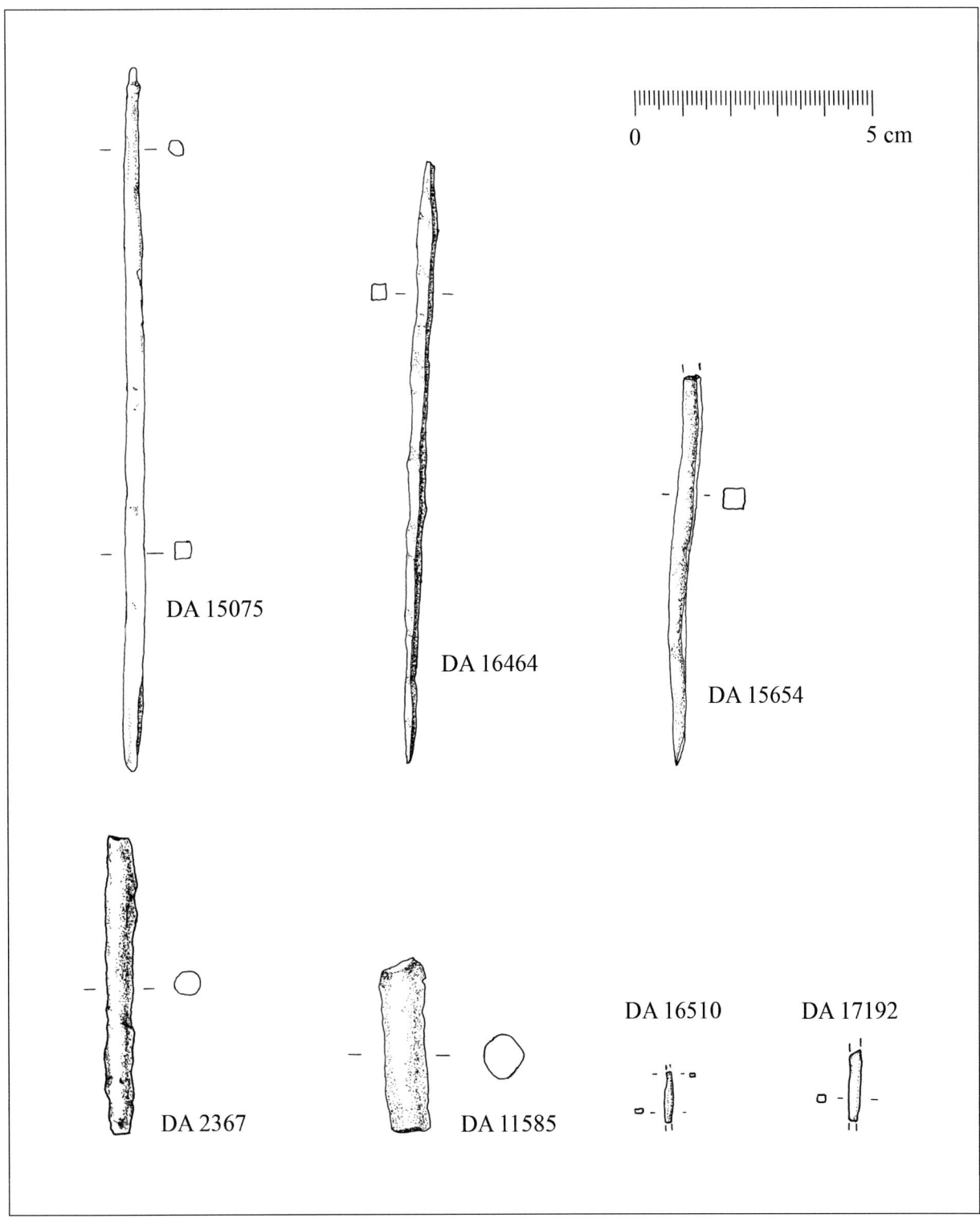

Figure 6.10. Ras Al-Hadd HD-6. Pins (drawings by L. Tricarico).

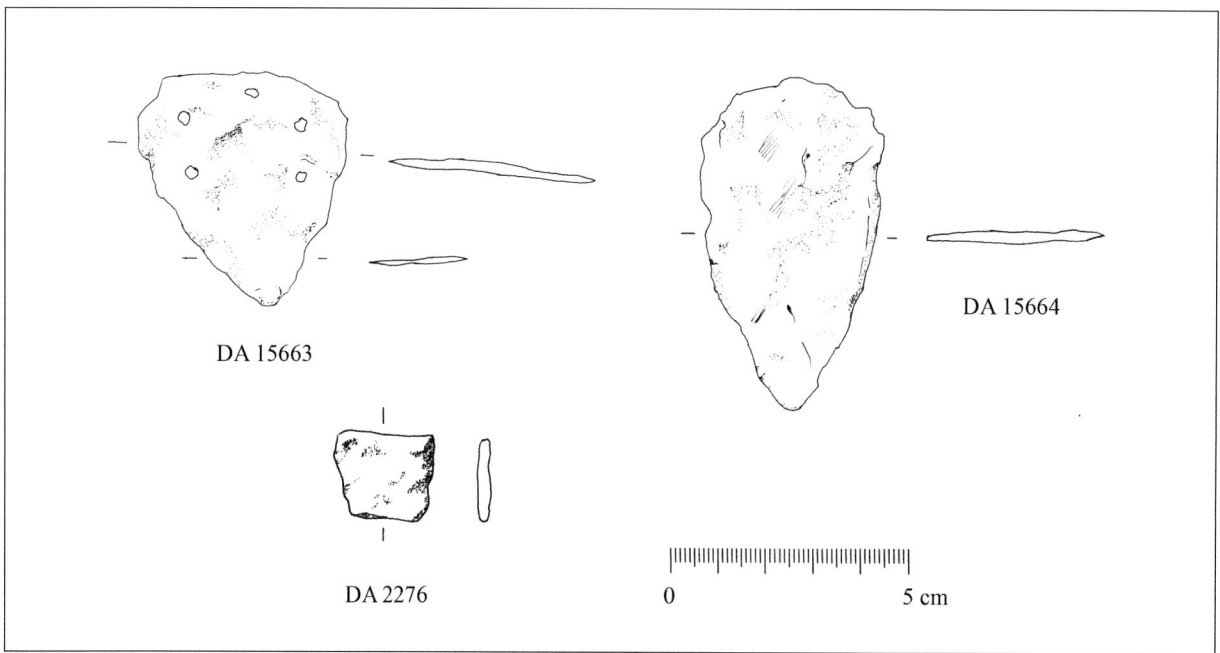

Figure 6.11. Ras Al-Hadd HD-6. Daggers (drawings by L. Tricarico).

The compositional analyses show an uncommon high arsenic content, up to 7.5% As, for item DA 16464. The use of an arsenic-copper alloy for pins may have the purpose to give these ornaments a special, silvering color. Pin DA 11585 has an anomalous lead content, 2.12% that could be attributed to a different copper ore source.

[Examined items: DA 2367 (fragment); 15075; 15654; 16464; 16510 (fragment); 17192 (fragment). Uncertain attribution: 11585 (semi-finished pin)].

Weapons

Daggers. Some daggers were recovered from different areas of the excavation (Figures 6.11 and 6.12). The analyzed items have a semicircular or straight blade base. The blades are not symmetrical, with the exception of DA 15664. They can be differentiated by the number of rivets lined up along the butt, having from three to five rivet-holes; the butt of dagger DA 15664 is missing. Thin sectioned blade thinning towards the cutting edge. Length between 4.5 and 6.9 cm; width between 4.4 and 3.6 cm. It is possible that the blade of dagger DA 17756 was reshaped (maybe after breakage) in order to obtain a scraper or a cutter. All the analyzed daggers show that they were made of an arsenic-nickel copper alloy, containing up to 4% of Ni and As, as a result the presence of this alloy only in weapons suggests that its use was intentional, probably as the result of an ore selection process. Arsenic and nickel could improve strength and hardness of copper, and arsenic-nickel copper alloy is particularly effective for weapons (Cheng and Schwitter 1957: 351; Weeks 2004: 112-113).

[Examined items: DA 15082 (from PQ 112-112, US 537, Room XXXIII, point 4; 3 rivet holes; 2 rivets are still preserved; 4.5x3.5 cm); 15663 (from D5, US 750, Room XXXIII, point 21; 5 rivet-holes; 4.9 x 4.4 cm); 15664 (from N 113, US 753; 6.9x3.6 cm); 17756 (from the surface; 4 rivet-holes; 4.5 x 3.2 cm). Uncertain attribution: DA 2276 (blade fragment)].

The Early Bronze Age: the Hafit period, ca. 3200-2800 BC

Figure 6.12. Ras Al-Hadd HD-6. Daggers (photograph by C. Giardino).

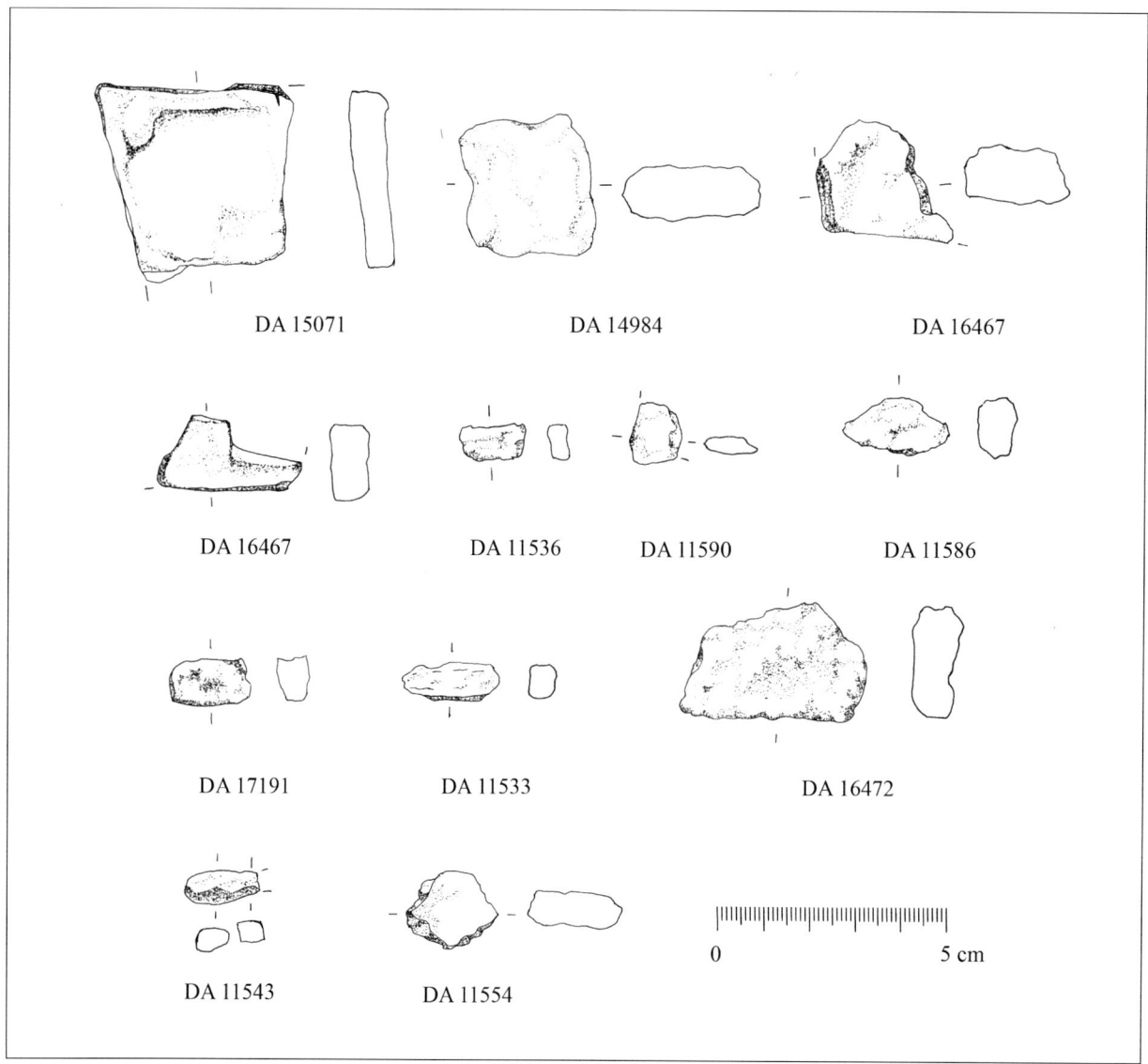

Figure 6.13. Ras Al-Hadd HD-6. Blocklets/flat pieces (drawings by L. Tricarico).

Semi-finished products and fragments for recycling

Blocklets/Flat pieces. They are semi-finished products typical of HD-6 metallurgical technology, which was based on cutting and hammering (Figure 6.13). Copper ingots were probably flattened in order to reach a thickness around 1 cm, then cut by chisel to grossly shape the object, obtaining a geometric blocklet; this semi-finished product was definitively shaped by hammering. Smaller scraps were sometimes collected together and tied with a rope (DA 11553). Blocklets are not characterized by a peculiar elementary composition, as we could expect from a semi-finished product, ready to be transformed into an object, most probably a tool. Most of them were made with almost pure copper; one of them (DA 14071) has been produced with an arsenic-nickel copper alloy containing 1.9% As and Ni.
[Examined items: DA 11533; 11536 (fragment); 11543 (fragment); 11554; 11586; 11590; 11596; 14071; 14984; 15071; 16465 (probably the beginnings of a fishing hook); 16467; 16472; 17191 (fragment)].

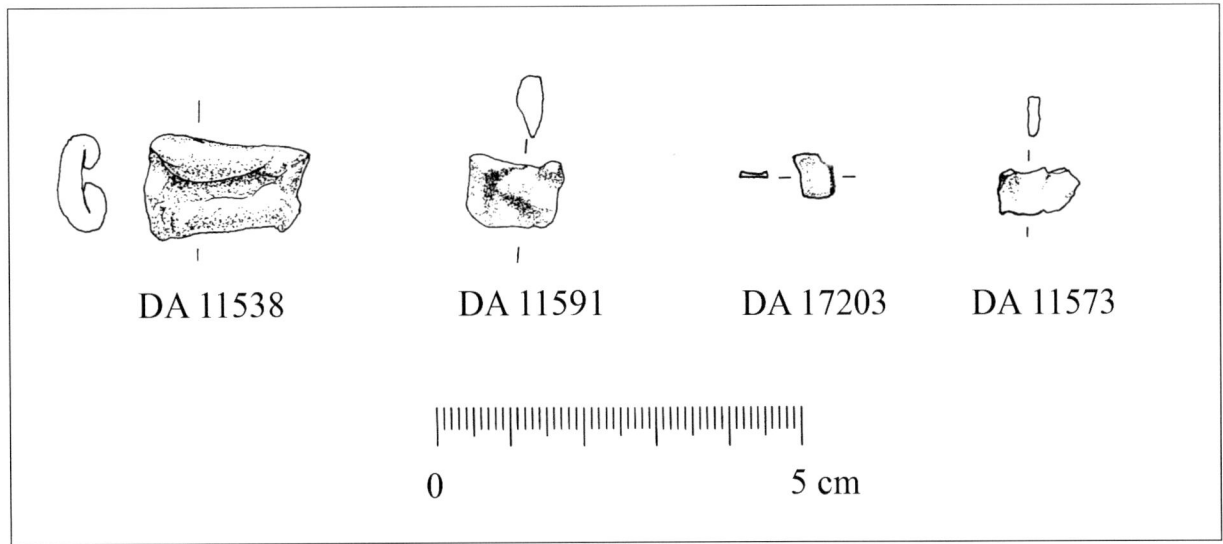

Figure 6.14. Ras Al-Hadd HD-6. Fragments for recycling (drawings by L. Tricarico).

Fragments for recycling. Small metal pieces occur at the site as fragments of broken objects – mostly sheet or blade fragments – destined to be recycled (Figure 6.14). Sometimes copper sheet fragments were folded. [Examined items: DA 11538 (sheet or blade fragment); 11573 (sheet or blade fragment); 11591; 14707/02; 17203 (sheet or blade fragment); 17805 (sheet or blade fragment)].

The metallurgical technology of Ras Al-Hadd HD-6

According to chemical analyses, copper used for the objects manufactured at HD-6 was smelted with rather primitive technologies. The chemical composition of the metalwork can provide an indication of the smelting process. The iron content of the objects is a good indicator of the smelting process used. In fact, primitive smelting is characterized by rather poor, reducing conditions, which are not sufficient to allow the iron from the ore to be incorporated into the forming copper. The situation changes completely in later slagging processes (Craddock and Meeks 1987; Craddock 1995: 137- 144). Therefore, copper smelted by a more primitive condition has a very low median iron content (about 0.03%), while the copper from more advanced techniques has Fe contents ten times higher (around 0.3%). The chemical analyses carried out on the objects from HD-6 reveal that the large majority has low iron percentages: they were made of rather pure copper, obtained with simple smelting processes, probably operating only within the limits of feasibility. The high nickel content is an indication that most probably copper came from the rich Ophiolite copper ore deposits distributed between the Omani mountainous chains and the island of Masirah.

No casting evidence was found at Ras Al-Hadd HD-6. On the other hand, the metallurgical investigations revealed that all the objects were simply made by mechanical deformation, using a working technology still largely reminiscent of Neolithic traditions. The metallographic structure of a semi-finished object shows a characteristic distorted dendrite structure, which indicates that the metal was cold worked after casting. Finished objects have a different structure, with twinned grains, revealing that the pieces were annealed after cold working.

Figure 6.15. Ras Al-Hadd HD-6. Blocklet DA 16467. Distorted dendrites produced by cold hammering (SEM image).

Figure 6.16. Ras Al-Hadd HD-6. Semi-finished pin DA 11585. Distorted dendrites produced by cold hammering (micrograph 5x).

Figure 6.17. Ras Al-Hadd HD-6. Awl DA 15684.02. Twinned grains after hammering and annealing (micrograph 200x).

The analytic evidence allows us to reconstruct the operative chain that took place in the fishing village of HD-6. Not only finished objects were recovered at the site, but also many copper pieces still in manufacturing progress: this is very helpful to examine the technological process. Copper ingots were imported as raw material to the village thanks to the large network of allegiances established by the tribes. The large amount of sea products coming from fishing was a relevant commodity to be used in these trades. In the village, the local artisans – working part time from their homes – produced tools, weapons and ornaments made with copper alloys. The ingots were first flattened by hammering to form thick copper plates (Figure 6.15). The plates were then cut in order to obtain all the different objects they produced.

Cutting was probably made with strong and solid chisels, as many such tools collected at the site during the excavation indicate. Then the piece was partially shaped by hammering (Figure 6.16). The craftsmen achieved the final object by cold and hot working, in other words, using an annealing technique (Figure 6.17). Annealing is a heat treatment that changes the physical properties of metal, increasing its ductility and making it more workable. After intense cold working, the object is heated in an oven for a while at high temperature, generally until glowing red; then it is left to cool. Annealing induces ductility, relieves internal stresses, refines the structure by making it homogeneous, and improves cold working properties; it prepares the item for further hammering processes. Hammering was not only useful for shaping an object, but also to increase its strength. After the production, the final object was submitted to the finishing process, where it was polished in order to acquire the bright reddish color peculiar of copper-based artefacts, and sharpened, in the case of tools or weapons, with a cutting edge. This final process was carried out with grained stones that were able to smooth the surface. Whetstones were found at HD-6, such as a fine-grained black sandstone whose surface had use-wear (DA 16471) (Figure 6.18).

Figure 6.18. Ras Al-Hadd HD-6. Whetstone DA 16471 (photograph by C. Giardino).

Many other stone tools were recovered at HD-6 during the excavation (Figure 6.19). Hard rocks were commonly used in Hafit metallurgical manufacturing as hammers and anvils: metal artefacts were in fact shaped on the flat surface of small, solid blocks, very similar to millstones in their aspect. To perform this operative chain specialised and skilled technicians were not necessary. Metallurgical production was a domestic industry, according to the archaeological evidence that attests that a metal working activity was found inside the houses of the village (Figure 6.20).

Probably, the families produced at home what they needed for the everyday life, such as fishing hooks, chisels, shell-openers and netting needles. Most likely, people from HD-6 were so adept at pyrotechnologies that they applied these techniques to other materials as well. Thousands of unglazed stone beads, different in shape and size, were recovered at HD-6 during the excavation, some of them made of steatite, also known as soapstone, a very soft talc-schist mostly composed by mineral talc, talc having one of the lowest definitional values on the Mohs hardness scale. The archaeometric analyses carried out by X-Ray Diffraction (XRD) allowed reconstructing their complex manufacturing process. They were carved out of solid, soft steatite, that was then hardened by firing at about 1000 °C for several hours, probably inside the same furnaces they used for copper annealing. Hardening is due to the transformation of steatite into synthetic enstatite, a transformation that takes place at 1000 °C. Similar beads were also found in Oman in graves from the oasis of Samad al Shan, in the Arabian Peninsula at Jebel Al-Emalah (UAE) and in the Umm an-Nar island; but also in Egypt, in Galilee and in the Indus Valley at Harappan sites (Panei *et al.* 2005).

There is a strong similarity between the two manufacturing processes, copper annealing and steatite hardening. At HD-6, metal and stone handcrafts were small-scale industries carried out at home by the families using their own equipment. Arguably, both working processes were carried out by the same people in the same structures, applying similar technologies. Of course an intriguing question should be if it was metallurgical knowledge that prompted to try the high temperatures to harden the soapstone, or if it was the reverse. Unfortunately, our data are still not enough to give a satisfactory answer. Nevertheless, the presence of stone manufacturing with fire hardening confirms the strong character of a cottage industry that characterizes all metal production at the site.

Figure 6.19. Ras Al-Hadd HD-6. Stone-hammer DA 23106 (photograph by C. Giardino).

The Early Bronze Age: the Hafit period, ca. 3200-2800 BC

Figure 6.20. Ras Al-Hadd HD-6. Reconstruction of the operative chain (drawings by L. Tricarico).

Figure 6.21. Ras Al-Hadd HD-10. Long dagger with straight blade DA 14334 (photograph by C. Giardino, modified by P. Koch).

Metal implements from other Hafit sites

Almost contemporary of Ras Al-Hadd HD-6 is the first period of the settlement of Hili 8 at Al Ain, in the Emirate of Abu Dhabi (UAE). This site is located at about 100 km from the sea and it is similar in size and several other aspects to the coastal site of HD-6, showing the complementarity that united the coast and the interior in technology and socio-economic life (Cleuziou and Tosi 2007: 93, fig. 101). The metal finds from Hili 8 (Period I) are small tools, like awls, weapons like daggers and ornaments like pins. Of course, the metal products related to fishing, that are so important at HD-6, did not exist here, in the interior of the country. The importance of these artefacts for the coastal fishermen is stressed by the burial furniture found in the collective graves of Ras Al-Hadd HD-10. This graveyard located on a cliff overlooking the settlement of HD-6 is also dated to the first centuries of the 3rd millennium BC.

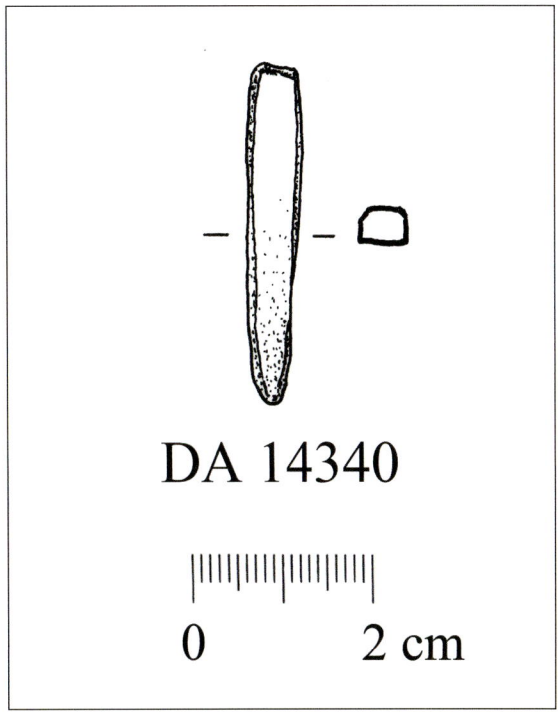

Figure 6.22. Ras Al-Hadd HD-10. Awl DA 10979 (drawing by L. Tricarico).

Here a fishhook was recovered, together with an awl and a long dagger with straight blade (Figures 6.21 and 6.22); two rivets joined the blade to the handle, made of organic material (perhaps wood) and now disappeared. The analyses show that blade and rivets were made of almost pure copper, although the rivets are heavily corroded. The blade also had a very high nickel content: 2% Ni. The awl was also made of almost pure copper, but rich in silver: 2.4% Ag. A similar silver content was detected on two small copper fragments belonging to a corroded, unidentifiable object.

About two hundred Hafit type cairn burials were found within a 2-km range in the area of Ras Al-Jinz – the easternmost cape of Arabia – grouped in small cemeteries of 10-30 collective graves. Ras Al-Jinz RJ-6 is one of these cemeteries, located at the foot of hills, dating from around 3000 BC. Ubaid painted pottery imported from Mesopotamia was recovered at this graveyard (Cleuziou and Tosi 2007: 92, 115, fig. 95).

Figure 6.23. Ras Al-Jinz RJ-6. Long pin with small flatted head DA 8621 (photograph by C. Giardino).

A long pin with a small flattened head (DA 8621: length 16 cm) was found in Tomb 1 (Figure 6.23); it has a chemical composition similar to other Hafit items because of high nickel content: it was a copper-nickel-arsenic alloy, with 95.50% Cu, 1.70% As, 1.28% Ni. The awl was also made of almost pure copper, but rich in silver: 2.4% Ag. A similar silver content was detected in two small copper fragments belonging to corroded, unidentifiable objects.

Other relevant documents of early metallurgy come from the area of the Bat oasis, on the southwestern slopes of the Al-Hajar Mountains. The excavations of the Bronze Age towers brought to light evidence of copper production. An uncommon copper platy slag was recovered together with copper ore at Kasr Al-Khafaji (Tower 1146), in a radiocarbon context dated ca. 3000 BC: they indicate that some form of small- scale copper processing occurred at the site (Thornton 2016). Metallurgical crucibles were recovered at Matariya (Tower 1147) and at Tower 1156 in Hafit / Umm an-Nar transitional contexts radiocarbon dated in both places to ca. 2900-2650 cal. BC. They were made of a chaffy ceramic and they are a very early evidence of pottery production devoted to metallurgical technology (Thornton and Ghazal 2016). At Tower 1147, prills, slag and small pins were also recovered together with crucible fragments, an indication that an early copper melting activity was carried out at this site for the production of metal objects (Leigh 2016). The Hafit copper melting evidence from Bat suggests that the oasis played a pivotal role in the development of metallurgy in Oman.

Chapter 7
Early Bronze Age: the Umm an-Nar period, ca. 2800-2000 BC

The term Umm an-Nar culture comes from the name of a site in the United Arab Emirates, an island ca. 15 km north-east of Abu Dhabi, where this kind of prehistoric assemblage was identified for the first time. Settlement and graves of this island show a characteristic, specialized material culture, characterized by strong economical adaptation to the local sea resources as fishing, dugong and turtle hunting, shell gathering. The Umm an-Nar culture identifies the Southeastern Arabian archaeological assemblages with tower-settlements, black-painted red ware, mortar-less stone masonry and a special type of cairns, dated around the second half of the 3rd millennium BC (Tosi 1976: 81).

The towers started to appear in Oman during the early Umm an-Nar period, ca. 2700-2400 BC and it seems probable that they were abandoned before the following Wadi Suq period. They are monumental structures, mostly made with stones, generally circular, with a diameter of about 20-25 m; towers occur in more than forty sites of the 3rd millennium BC, as Hili, Bat, Al-Moyassar, Ibra, etc. The towers testify the complexity of the Omani social ties in the Umm an-Nar period (Cable and Thornton 2013; Thornton et al. 2013). Metallurgical activities occurred inside the towers. At Bat, evidence of copper refining and, perhaps, of smelting was recovered at Tower 1156, constructed during the early Umm an-Nar period. The presence of a little slag in the Umm an-Nar levels, together with fire pits associated with industrial activity and copper prills, suggest that refining or, perhaps, smelting processes of high-grade copper oxide ores took place at the site, where there are also clues of interaction with the Indus Valley (Mortimer 2016).

One of the most characteristic features of this culture are its peculiar burials. During the second quarter of the 3rd millennium BC, a new type of cairn was erected, the Umm an-Nar type graves, still circular, but with the interior divided into chambers by stone sections, in order to place more bodies inside. The smallest ones are the most ancient, while the larger belongs to the end of the 3rd millennium BC (Figure 7.1). This type of tomb was therefore in use for about seven centuries, from ca. 2700 BC until ca. 2000 BC. They were now erected closer to the settlements. Two small doors frequently closed the graves, because the monuments were divided into two non-communicating parts. A distinctive element is a plinth of flat slabs protruding from the external walls. The outside is generally elegantly worked with the stone carefully shaped by hammering.

The Umm an-Nar culture developed in a later phase of the Early Bronze Age, a relevant period in the evolution of Arabian metallurgy. In this phase tin bronzes started to appear in Southeastern Arabia, even though they were very rarely used and mostly were for prestigious items. The addition of tin lowers the copper melting point; it increases also the casting fluidity and improves mechanical properties such as malleability and toughness. It modifies the color of copper too, giving to the metal a less reddish and more golden shade.

The first metallurgical workshops that used a casting process for making artefacts dates to this period. Early evidence of copper castings was recovered at Ras Al-Jinz RJ-2, a coastal fishing site in the Ja'alan region (Figure 7.2). It should be stressed that until now, this is one of the earliest indications of melting activities in Oman. Crucible fragments containing bronze residues were discovered in the same place: this is a clear indication that at that site a specialized activity was carried out such as casting tin copper alloys.

Figure 7.1. Umm an-Nar cairn burial at Bat (photograph by C. Giardino).

Figure 7.2. Ras Al-Jinz, the area of RJ-2 (photograph by C. Giardino).

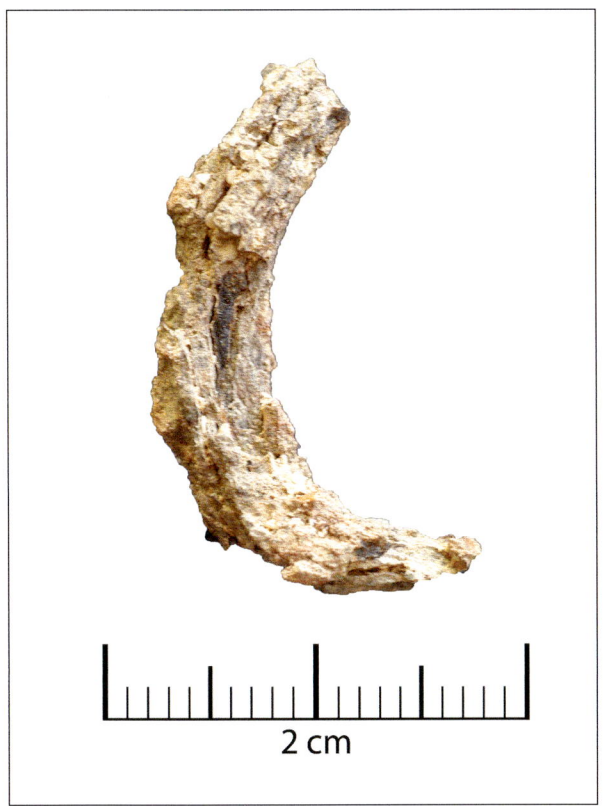

Figure 7.3. Ras Al-Jinz RJ-2. Iron fishhook (DA 10958) (photograph by C. Giardino).

The first evidence of iron comes from the same site, a fragmentary fishhook 2.5 cm long, DA 10958 (Figure 7.3). It was analyzed by XRF in several points, also in the core (the hook is broken), in order to reject the hypothesis that it could be a copper object with only a natural iron patination. The elementary composition is: 92.6% Fe; 7.4 Ca (due to calcareous concretions). Nickel was not detected in this item. This is an indication that the hook was not made of meteoric iron, because ordinarily iron meteorites contain about 5-20% nickel (Giardino 2010: 196).

At Umm an-Nar island, in the Emirates, very close to Oman, brass (zinc-copper alloy), one of the most elusive alloys of the Bronze Age, was discovered in a grave furniture. A dagger from this site contained 10% Zn, while a fragment has 8.6% Zn (Frifelt 1991: 100; Thornton 2007). The dagger was a relevant status symbol in this period: the golden hue of brass made it immediately distinct from other copper alloys.

Both metals, iron and brass require specific processing techniques and a deliberate investment of time and labor. Their presence in Umm an-Nar contexts indicates a high level of working knowledge of materials and material processing on the part of the Arabian metallurgists (Thornton and Ehlers 2003).

Ras Al-Jinz RJ-2: a metallurgical workshop

The Ras Al-Jinz cape is the eastern most corner the Arabian Peninsula, located 11 km south of Ras Al-Hadd. Immediately south of the cape, the cliffs are cut by the ocean and are bordered with a sandy beach more than one kilometer long, where the big green turtles nest. In the middle of the beach stands an isolated rise with a flattened top, where Iron Age buildings were erected, while the Bronze Age sites were located at the foot of its northern side (Coltorti 1989).

Figure 7.4. Ras Al-Hadd HD-5 (surface). Copper seal (DA 12803) (photograph by P. Koch, courtesy Ministry of Heritage and Culture, Sultanate of Oman).

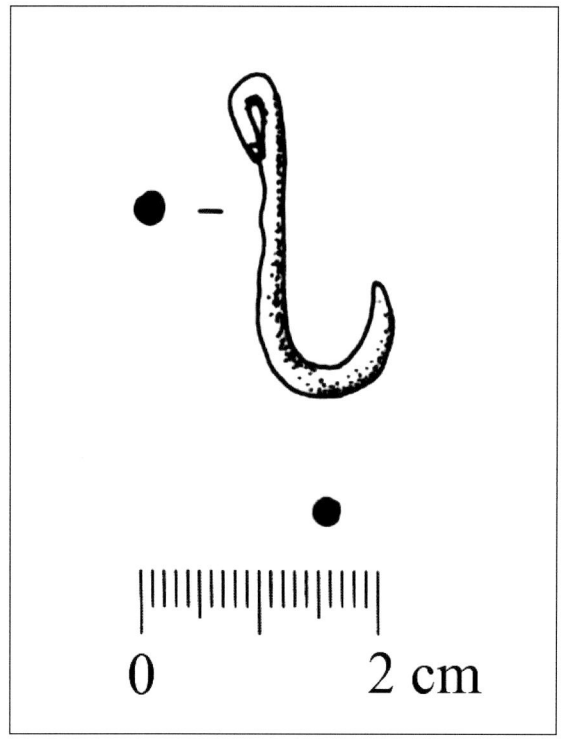

Figure 7.5. Ras Al-Jinz RJ-2. Harappan style fishhook (DA 12792).

A reference point for seafaring in the Arabian Sea is the little mountain of the Jebel Salim Khamis (325 m) that lies in the background of Ras Al-Jinz, about 40 km from the beach: this landmark was the first sign of the coasts of the Arabian Peninsula for the sailors coming from India. The location is extremely favorable for fishing and navigation. Therefore, it was settled from the Late Holocene to the Islamic period. Both settlement (RJ-2) and necropolises (RJ-6, RJ-10, RJ-13 and RJ-20) can be dated to the beginning of the 3rd millennium BC. The tombs are collective graves like the Hafit cairn burials at RJ-6, even though they differ in shape and architecture. The economy of RJ-2 was based on the intensive exploitation of marine resources that were traded in the interior of the country. The fish was salted, smoked and dried, while the shells were processed locally to produce rings and pendants, a kind of ornament spread across Oman. The trade with the other side of the Indian Ocean was flourishing: pottery sherds with inscriptions from the Indus Valley were recovered on site.

Another relevant find was a copper-based seal with signs of the Indus script (Tosi 1998: 172, 234, fig. 176; Cleuziou and Tosi 2000: 59- 60, fig. 17). That seal has not yet been analyzed. Another square seal (DA 12803) was collected at Ras Al-Hadd HD-5; it was similar in shape, but with geometric motives (Figure 7.4). This one was examined by XRF; the metal was almost pure copper (97.5%), in order to ease the engraving process that was employed for the decoration on its surface: a dot in the center surrounded by three concentric diamonds and four dots in the corners of the seal. In fact, the parallel lines of the diamonds were made by carving the metal with a chisel; the dots in the corners and in the center were produced by punching. The strong level of interactions with India is testified by the presence at Ras Al-Jinz RJ-2 not only of prestigious items, but also of everyday tools, as a fishhook with an eye belonging to Harappan types (Figure 7.5). Nevertheless, the site was also connected with Mesopotamia, as testified by many jar fragments of Mesopotamian origin recovered at RJ-2, together with hundreds of bitumen pieces, coming most probably from central Iraq (Cleuziou and Tosi 2000: 44-53).

More than 1500 metal objects were recovered during the excavations. The metal was worked locally, in special workshops; the raw material was imported from centers in the interior of the country, together with other supplies, such as dates, in return for sea products. The life of RJ-2 started with an aceramic phase (Period I: 4th millennium BC); then, after an abandonment gap, a new settlement was created in the Early Bronze Age, with the construction of mud-brick buildings. Period II is dated, also according to Harappan material from the site, between 2500 and 2300 BC; Period III between 2300 and 2100 BC, according to radiocarbon dates, while Period IV falls around 2000-2100 BC (Figure 7.6) (Cleuziou and Tosi 2000; Giardino and Lazzari 2014). The activity of the metallurgical workshops developed during the second half of the 3rd millennium BC. Hammering using hard rock tools as hammer-stones and anvils shaped most of the copper objects; engravings were produced with metal chisels.

The Early Bronze Age: the Umm an-Nar period, ca. 2800-2000 BC

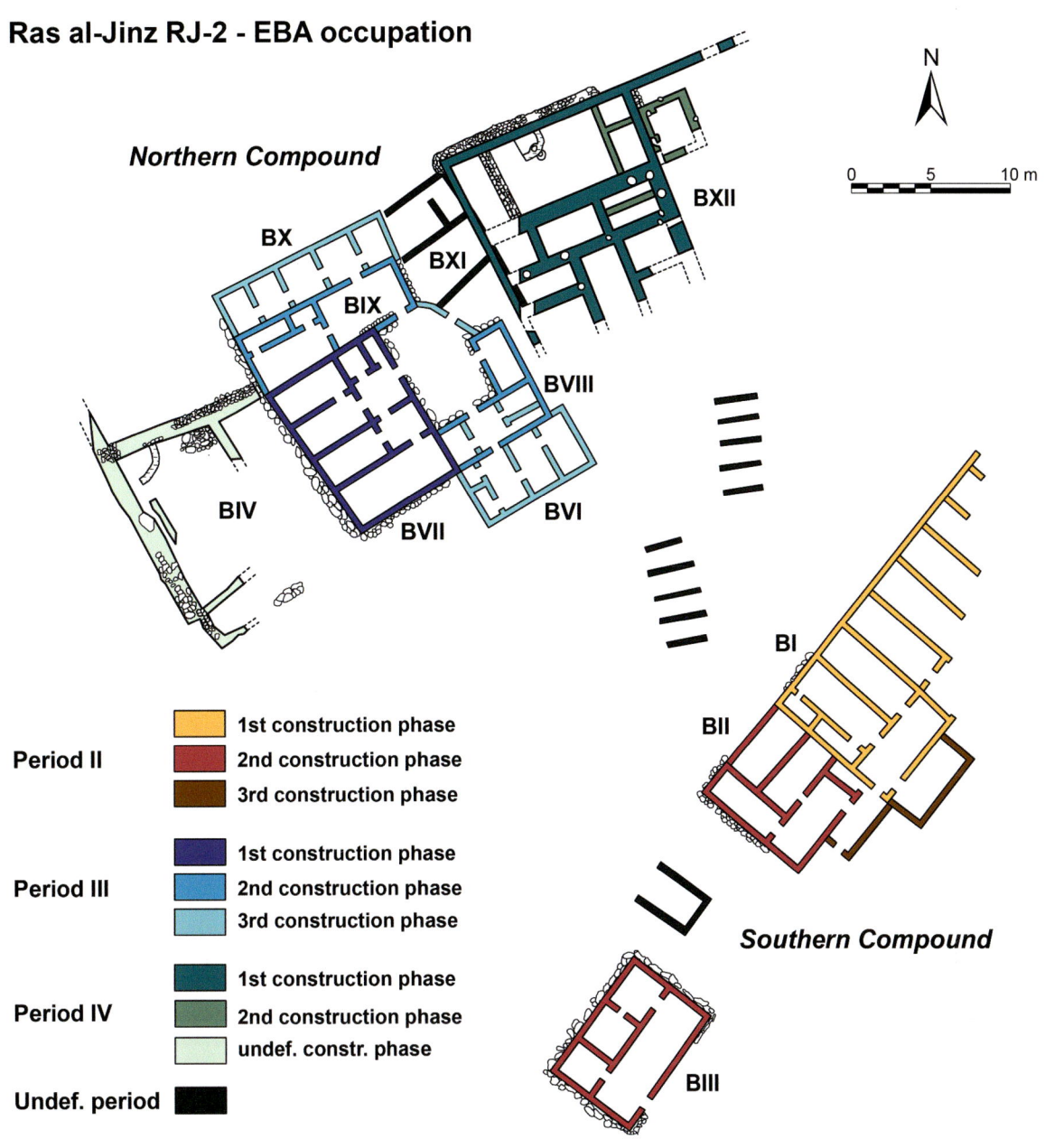

Figure 7.6. Ras Al-Jinz RJ-2. Prehistoric structures and their chronology (image by V. Azzarà, courtesy Ras Al-Jinz Study Program).

A large fragment of a flat ingot from Building I, Room 4 (DA 8506), has a line carved by chisel on its surface (Figure 7.7). Probably that line marks the point where the metal was to be cut to obtain a semi-finished piece devoted to become the final object. A flat copper bar from the same Building I, Room 2 (DA 8366), displays cut edges and it has the same size of the semi-finished piece traced on the flat ingot (Figure 7.8). Some remains of fireplaces recovered during the excavation could be connected with metallurgical activities as casting or annealing. Some crucible fragments (Figures 7.9 and 7.10), together with a small casting residue, were found in different areas of the settlement (Figure 7.11), as in Building IX, Room 9 and east of the Northern Compound (Cleuziou and Tosi 2000: 56-57): they are clear evidence that casting was carried out at the site, in different workshops.

Figure 7.7. Ras Al-Jinz RJ-2. Flat ingot (DA 8506). Arrows mark the engraved line on its surface (photograph by C. Giardino, modified by P. Koch).

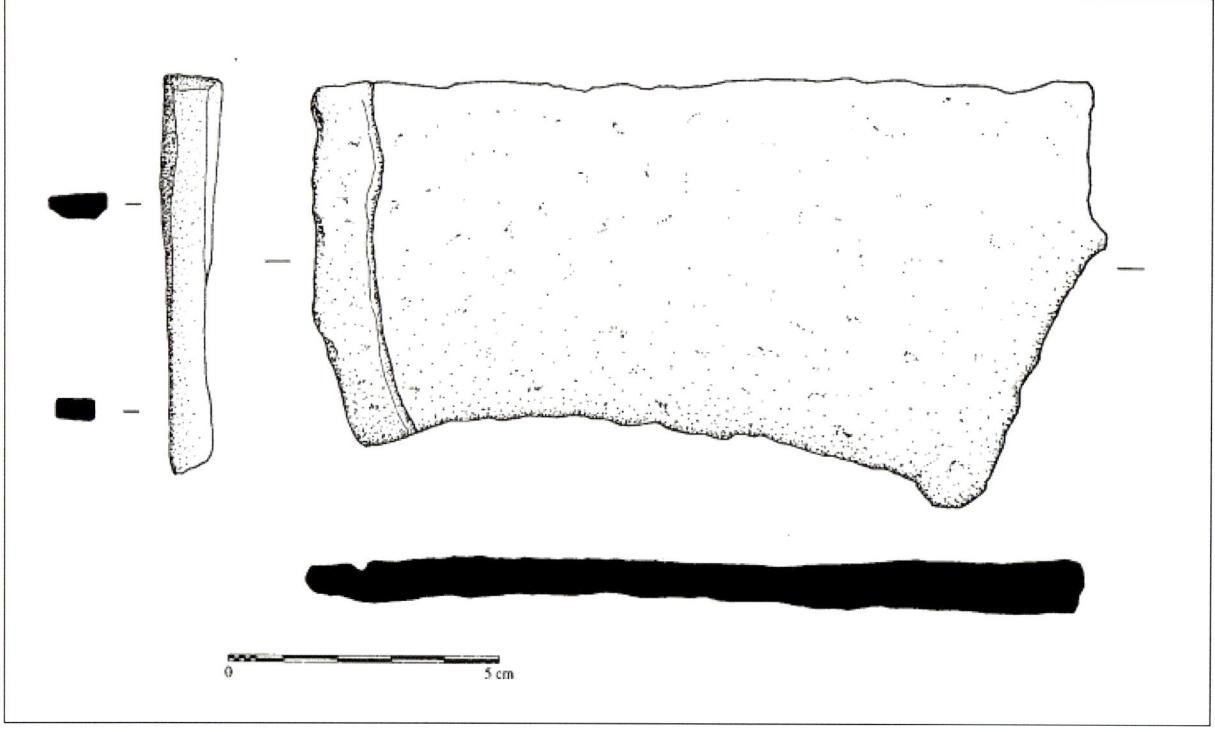

Figure 7.8. Ras Al-Jinz RJ-2. Flat ingot Building I, Room 4 (DA 8506) and flat copper bar from Building I, Room 2 (DA 8366).

The inner side of the crucibles was partially vitrified by the high temperature of the molten copper alloys that were contained inside, around 1000 °C. Pouring molten metal inside a pre-shaped form allowed to produce in a simple and quick way objects of complex shapes. Crucibles were made of refractory clay, different from ordinary pots. The presence of foundry remains at RJ-2 is an indication that the metallurgical activity was now carried out by specialized artisans that were able to cast the metal to make artefacts and perhaps to alloy copper with tin as well.

The fragments were analyzed by XRF, in order to detect which kind of metal was poured inside. Some reveal the presence of high contents of copper oxide, associated with lead and traces of nickel. Other crucibles show a relevant tin content, with copper, iron, zinc, arsenic (Giardino and Lazzari 2014) (Figure 7.12). No molds were recovered during the excavation at RJ-2: this is not very surprising, considering that molds are a rare find also in metallurgical archaeological contexts and that they are under-represented in the archaeological record.

Figure 7.9. Ras Al-Jinz RJ-2. Crucible fragments (RJ-2: 551,30-551,31). One crucible has its inner surface vitrified and green colored by copper (photographs by C. Giardino, modified by P. Koch).

It is possible that here, as in other geographical areas, molds were made of perishable materials, such as clay or sand, that disintegrate after casting, leaving no archaeological evidence (Ottaway and Seibel 1998; Ottaway and Roberts 2008: 212). The presence of relevant technological innovations – casting and the introduction of bronze alloys – does not mean the total disappearance or the withdrawal of the older techniques. There is evidence that many objects were still made by cutting flattened ingots, and then shaped by cycles of cold working and annealing in order to achieve a sufficient deformation, according to the Hafit tradition testified at HD-6. It is possible that a specialized metallurgical activity connected with the new technologies coexisted with the old cottage industry, carried out to support basic, domestic needs.

Figure 7.10. Ras Al-Jinz RJ-2. Crucible fragments (RJ-2: 551,32-551,37). The inner surface of the crucible is badly damaged and vitrified by the high temperature of molten metal (photographs by C. Giardino, modified by P. Koch).

Types of artefacts from Ras Al-Jinz RJ-2

Most of the metalwork found at Ras Al-Jinz RJ-2 is similar to the items already described in the Hafit settlement of Ras Al-Hadd HD-6: the two sites share a common economy devoted to the exploitation of sea resources. Therefore, many specialized tools as fishhooks, awls, netting needles and shell-openers are almost identical in both places. Particularly, the fishhooks are one of the most common metal items found at Ras Al-Jinz RJ-2: they have a similar shape, but with slight differences, that was probably connected with the various methods for catching fish (Figure 7.13).

Nevertheless, the metallurgy is clearly more developed at Ras Al-Jinz RJ-2 than at Ras Al-Hadd HD-6. Casting activities took place in the village, as shown by the presence of crucible fragments. According to XRF analyses carried out on the crucibles, not only pure copper or arsenical copper was used for the production of metal objects, but also tin bronze. Another relevant indication of technological development is the appearance of new shapes and types, like flat axes.

Figure 7.11. Ras Al-Jinz RJ-2. Small casting residue (DA 12673) (photograph by C. Giardino, modified by P. Koch).

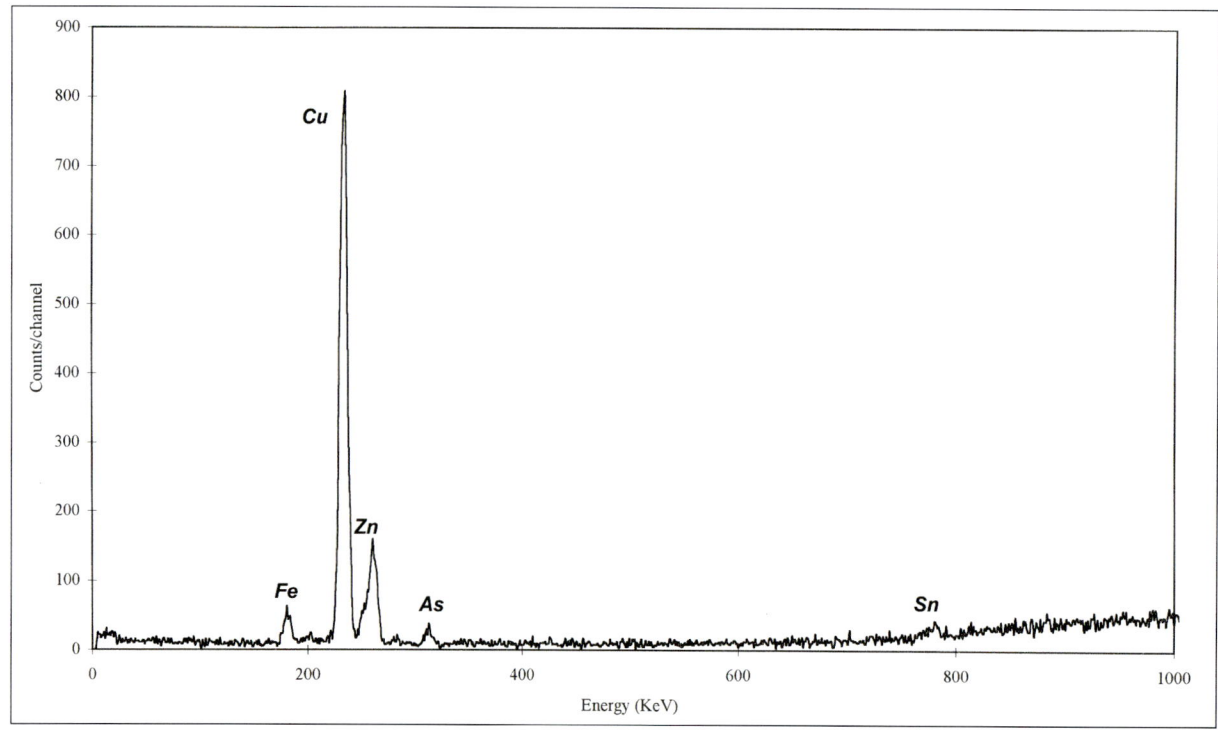

Figure 7.12. Ras Al-Jinz RJ-2. ED-XRF spectrum from a crucible fragment (from QDF, U.S. 6062) containing a tin-copper alloy.

The Early Bronze Age: the Umm an-Nar period, ca. 2800-2000 BC

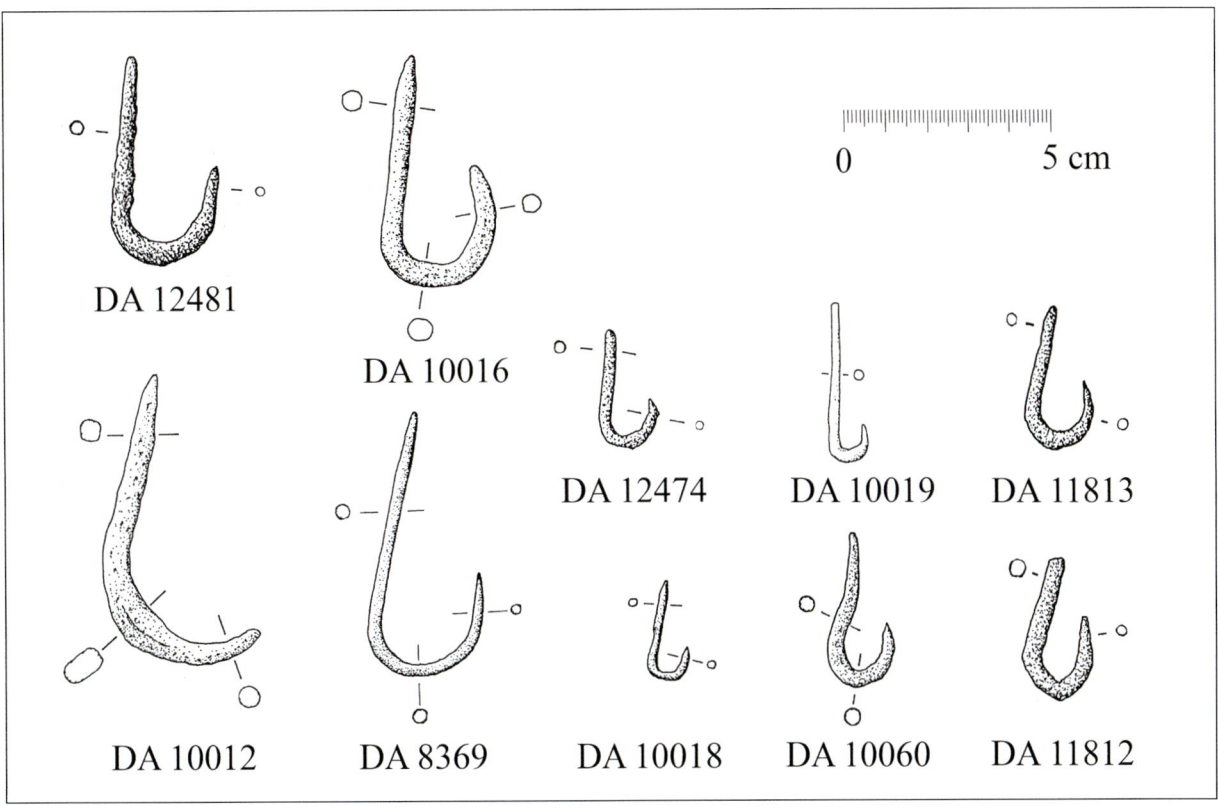

Figure 7.13. Different fishhooks from Ras Al-Jinz RJ-2.

Figure 7.14. Ras Al-Jinz RJ-2. Fishhooks (photographs by C. Giardino, modified by P. Koch).

Figure 7.15. Ras Al-Jinz RJ-2. Awls (photographs by C. Giardino, modified by P. Koch).

Tools

Fishhooks. Fishhooks were common in the settlement (Figure 7.14). They differ in size, but always have straight shanks, like the more ancient specimens from Ras Al-Hadd HD-6. The line was twisted and tied to the shank: DA 8335 still has traces of the mineralized line. An exception is hook DA 12792, with the end of the shank bent over to form an eye; this peculiar shape is common in the Bronze Age Indus Civilization (Chakrabarti and Lahiri 1996: 39, fig. 1). The object can be regarded as an import from India. This hook does not differ from other Omani hooks for its composition: it is characterized by little arsenic and nickel contents (0.28% As; 0.18% Ni). Hooks were made of almost pure copper, with an arsenic content less than 2%. Another completely unusual item is DA 10959, made of iron instead of a copper alloy.

[Examined items: DA 8335 (bent; trace of the mineralized line around the shank); 8336 (fragment); 8703 (unfinished hook?); 8707; 10012; 10038 (fragment); 10057 (unfinished hook?); 11999 (unfinished hook?); 12474; 12479 (unfinished hook); 12774; 12785; 12897 (fragment). Exceptional items: DA 12792 (from square LVA-LVB – US: 5) (Harappan style); 10959 (iron)].

Awls. Awls from RJ-2 are similar, for shape and function, to the same tools from HD-6 (Figure 7.15). Some of the awls have a remarkable arsenic content, up to 3.3% As, frequently associated with a high nickel percentage, up to 2.7% Ni. Two items show some tin presence, up to 2.4% Sn. It is possible that they were the recycling product of old bronze objects.

Figure 7.16. Ras Al-Jinz RJ-2. Chisel (photographs by C. Giardino, modified by P. Koch).

[Examined items: DA 8367 (flat); 8424; 8459; 8577; 8712 (fragmentary); 10853; 12362 (fragmentary); 12495; 12741; 12906].

Chisels/Flat chisels. A chisel is a tool with a characteristically shaped cutting edge blade on its end, for carving or cutting hard materials; some of the RJ-2 chisels are flat and thick, with one or two cutting edges (Figure 7.16). Chisels were produced with pure copper, but also in arsenic-copper alloy, where arsenic reaches a content of 3.6% As.
[Examined items: DA 8687; 10043; 10062; 10186; 12548; 12734; 12739; 12742].

Figure 7.17. Ras Al-Jinz RJ-2. Crochets/netting needles (photographs by C. Giardino, modified by P. Koch).

Figure 7.18. Ras Al-Jinz RJ-2. Shell-openers (photographs by C. Giardino, modified by P. Koch).

Drill points. Drill points are small copper-based pointed fragments for piercing hard materials. Only one drill point was analyzed: it was made of almost pure copper.
[Examined items: DA 10056].

Crochets/Netting needles. Small tools that were probably related to the production and restoration of fishnets (Figure 7.17). Two crochets were analyzed: they both had an arsenic content around 2.5% As.
[Examined items: DA 12089; 12504].

Shell-openers. This specialized tool is linked with the exploitation of edible shells (Figure 7.18). Some of these items could be produced by recycling broken awls, by hammering. They were made of pure copper and with arsenic-copper alloys, where the As is around 1-2%. One has a little tin content, 0.6%, probably related to a recycling process.
[Examined items: DA 10032; 12076; 12503; 12913; 8371].

Knives. They are small, tanged knives with straight blades (Figure 7.19). They could be used to cut nets or lines. The older fishing settlement of HD-6 seems to lack this kind of tool. The analyses show that they have a low arsenic content, less than 2%, associated with similar percentages of nickel.
[Examined items: DA 8574; 11982; 12357].

Flat axes. Axes can be regarded as an exceptional find in Umm an-Nar settlements (Figure 7.20). The unbroken specimen from RJ-2 is a narrow elongated, flat, trapezoidal axe with enlarged butt; cutting edge is almost straight. The two items were made using almost pure copper; the fragment # 1382 has 98.5% Cu, while DA 10064 shows contents of arsenic (1.8%) and nickel (1.1%).

The Early Bronze Age: the Umm an-Nar period, ca. 2800-2000 BC

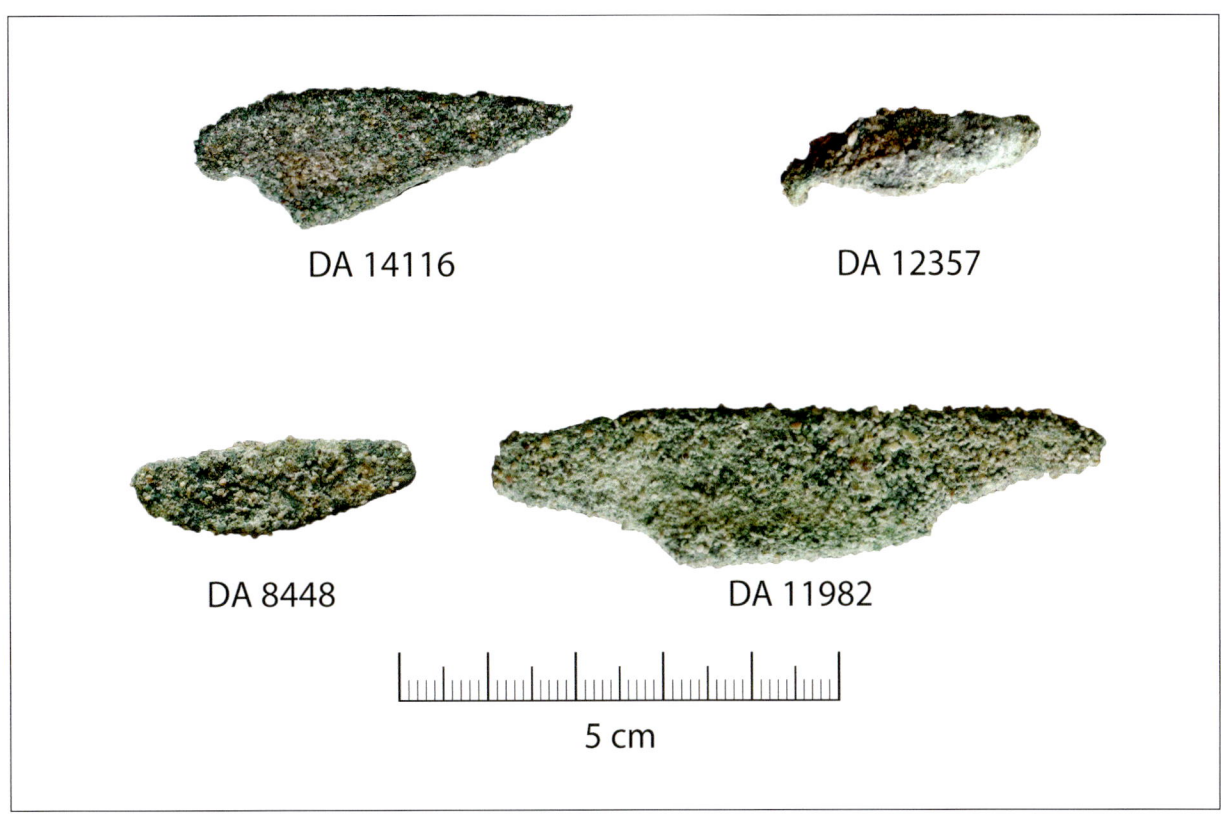

Figure 7.19. Ras Al-Jinz RJ-2. Knives (photographs by C. Giardino, modified by P. Koch).

Figure 7.20. Ras Al-Jinz RJ-2. Flat axes (photographs by C. Giardino, modified by P. Koch).

Figure 7.21. Ras Al-Jinz RJ-2. Pins (photographs by C. Giardino, modified by P. Koch).

These items have parallels with another axe from Al-Moyassar-1, which has a similar size (11.5 cm long) (Weisgerber 1980: fig. 75, 5; Weisgerber and Yule: 46-47). Flat axes are rather common in Umm an-Nar contexts, both in tombs and in settlements, but most of the items are longer than the sample from RJ-2, such as the axes from Umm an-Nar, Al-Moyassar-4 (Grave 3), Tell Abraq, ranging from 16 up to 30 cm. [Examined items: DA 10064 (14.5 x 4 cm; from Building II: XVI, 708); #1382 (blade fragment, with cutting edge: 4.1 x 2.5 cm; from the surface)].

Ornaments

Pins. Pins were generally used for fastening together pieces of cloth (Figure 7.21). The analyzed pins were made of almost pure copper.
[Examined items: DA 12497; 12666 (globe-headed, fragmentary); 12786].

Rings/Earrings. Rings with rounded sections, with closed, open or overlapping ends (Figure 7.22). The examined rings had low arsenic content, up to 2.2% As; one of them also had some tin: 1.3% Sn.
[Examined items: DA 8697 (open ends); 12750 (closed ends); 12775 (overlapping ends); 12791].

Razors. They were symmetrical razors with tanged handles; they have oval or rectangular blades with rounded lower edges (Figure 7.23). Two razors were made with bronze; one of them has a very high tin percentage, 16.4% Sn. This content shows that alloying was intentional and that razors were prestigious items.
[Examined items: DA 12732 (fragment); 12733 (fragment; the piece was bent in antiquity for recycling); 12736 (fragment)].

Beads. Many metal beads were found at RJ-2. They were biconical, globular, barrel-shaped or cylindrical. Particular importance has a necklace made of stones and copper based beads, oval shaped (DA 12109) (Figure 7.24). According to the analyses carried out by SEM-EDS, some of the metal beads were made with copper, some other with tin bronze; the stones were white calcite.
[Examined items: DA 12109, from Building VII, Room 2].

Semi-finished products and fragments for recycling

Blocklets/Flat pieces. Blocklets are typical items of a still primitive metallurgical activity, based on cutting and hammering, that survived beside a more developed casting technology. The items have mostly geometric shapes, rectangles, triangles or diamonds (Figure 7.25). The chemical composition reflects their nature of semi-finished products. Some of them contain around 1% arsenic; some others are composed of almost pure copper. One blocklet (DA 12735) shows an exceptional amount of arsenic (18.2% As), that is associated with a lot of lead (17.2% Pb), antimony (3% Sb) and nickel (2.6% Ni): it could have had a different origin from the other items.
[Examined items: DA 8366; 8438; 8477; 8495; 10055; 10944; 10949; 11997; 12487; 12488; 12489; 12490; 12491; 12492; 12493; 12735; 12738 (pierced); 12777; 12896].

Flat ingots. These semi-finished products were probably obtained by flattening imported copper ingots, in order to obtain objects by cutting and hammering (Figure 7.26). An intriguing evidence of this technique is the engraved line on the surface of item DA 8506. The line, carved by chisel, marks the precise point where the flat ingot had to be cut in order to produce the desired object. DA 8506 was made of a copper-based alloy containing mostly copper (97.5% Cu), with some arsenic (1.3% As) and nickel (0.7% Ni), a rather common composition of Omani prehistoric items.
[Examined items: DA 8506 (with an engraved line on the surface); 12507 (uncertain attribution: this item could also be a fragment of hoe)].

Casting residues. Fragments of residues dropped out from a crucible during the casting process. Together with crucible fragments, they are a relevant indication of copper casting in the settlement. Their composition is almost pure copper (99%), with only few traces of other elements.
[Examined items: DA 12086 (lump); 12673 (from QIV-QIY)].

Sheets or blades/Fragments for recycling. Not definable, small sheet fragments belonging to blades (razors?, daggers?) or metal vessels, probably for recycling (Figure 7.27). The analyses show arsenic up to 3.6% and nickel up to 1.9%.
[Examined items: DA 8374; 12508; 12509].

Figure 7.22. Ras Al-Jinz RJ-2. Rings/earrings (photographs by C. Giardino, modified by P. Koch).

Figure 7.23. Ras Al-Jinz RJ-2. Razors (photographs by C. Giardino, modified by P. Koch).

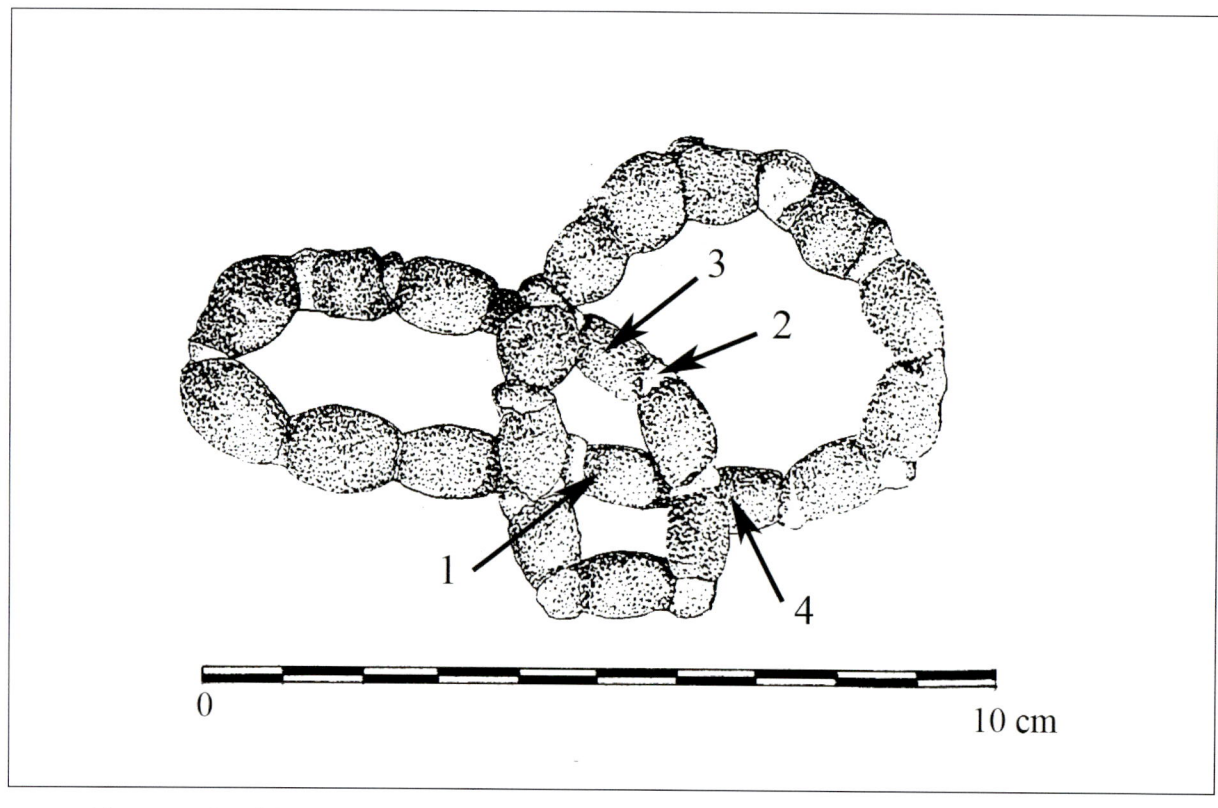

Figure 7.24. Ras Al-Jinz RJ-2. Necklace from Building VII with indication of the sampling spots for SEM/EDS analysis.

Figure 7.25. Ras Al-Jinz RJ-2. Blocklets/flat pieces (photographs by C. Giardino, modified by P. Koch).

Metals from Omani Umm an-Nar graves: Ras Al-Hadd HD-7

Early tin bronzes were discovered inside Umm an-Nar tombs from the coastal site of Ras Al-Hadd HD-7. The metal grave furniture comes from two burials, Tomb 4 and 5, dated from the second half of the 3rd millennium BC (Figure 7.28). Tomb 5 was particularly rich in metals: it contained a razor (DA 22596-5019), a ring (DA 22591-5015), a chisel (DA 22589-5013), an awl (DA 22595-5018), three pins (DA 22585-5009, DA 22586-5010 and DA 22587-5011), a small fragment of a possible ingot (DA 22588-5012) and a fragment of a plate (DA 22588-5012). Tomb 4 had only a razor (DA 22565-4031). According to the results of the XRF and SEM analyses, both razors were made of bronze, so as the ring, one pin and the plate fragment; tin content was between 1.2 and 4.4%. Tin was associated with arsenic: this association was observed also in other metal objects from Umm an-Nar graves, at Al-Sufouh, Unar 2 and Tell Abraq (Weeks 2003: 86, tabs. 4.1, 4.4, 4.4). Most of the items from HD-7 graves have a high nickel percentage, between 1 and 2%.

Umm an-Nar metallurgy: some remarks

Umm an-Nar metal technology is rather more developed than the previous Hafit metalworking. Evidence of castings was recovered at Ras Al-Jinz RJ-2, at Al-Moyassar and at Umm an-Nar island, a place where fragments of crucibles, molds and casting residues were recovered (Weisgerber 1983: 271; Frifelt 1995: 70, 188-191; Weeks 1997: 17-20).

Figure 7.26. Ras Al-Jinz RJ-2. Flat ingots (photographs by C. Giardino, modified by P. Koch).

Other copper workshops are recorded at Hili 8 (Periods IIe and IIf), Tell Abraq, Bat and Ghanadha (Weeks 2003: 54). The majority of metal items from the second half of the 3rd millennium BC was produced with almost pure copper; copper-based objects generally contained only small amounts of arsenic, up to 2-3%. Very few finds were made of the new alloy, tin bronze. The typology of the bronze objects shows the lack of interest for mechanical properties: rings, razors, beads were in fact ornament and prestige items. Working tools were frequently made with almost pure copper. An intriguing object is the necklace found almost intact at Ras Al-Jinz RJ-2, Room 2 of Building VII, made of copper-based oval shaped beads alternated with round, stone beads. The analyses, carried out by Scanning Electron Microscope (SEM) associated with EDS revealed that the metal beads were not made with the same alloy: some of them were made with copper, other with tin bronze, in order to give different colors to the necklace; the stones were calcite (Figure 7.29).

Figure 7.27. Ras Al-Jinz RJ-2. Sheet or blades: fragment for recycling (photographs by C. Giardino, modified by P. Koch).

Therefore, to understand the real aspect and value of this piece of jewelry we must consider the full palette of its colors. Originally, the necklace had a precious and colorful appearance, thanks to the bright, red copper, the golden bronze and the white stones. The use of different copper alloys in order to obtain special chromatic effects is an evidence that for the Umm an-Nar communities tin bronze was mostly devoted to the construction of visual effects related to light and brilliance. As in many prehistoric contexts, color was a powerful mean of objectifying social and symbolic values in the Bronze Age Omani society (Jones and MacGregor 2002: 12-13). Though tin bronze technology started to spread in the southeast part of the Arabian Peninsula from the Umm an-Nar period, there are no tin ore sources in the whole area.

The Early Bronze Age: the Umm an-Nar period, ca. 2800-2000 BC

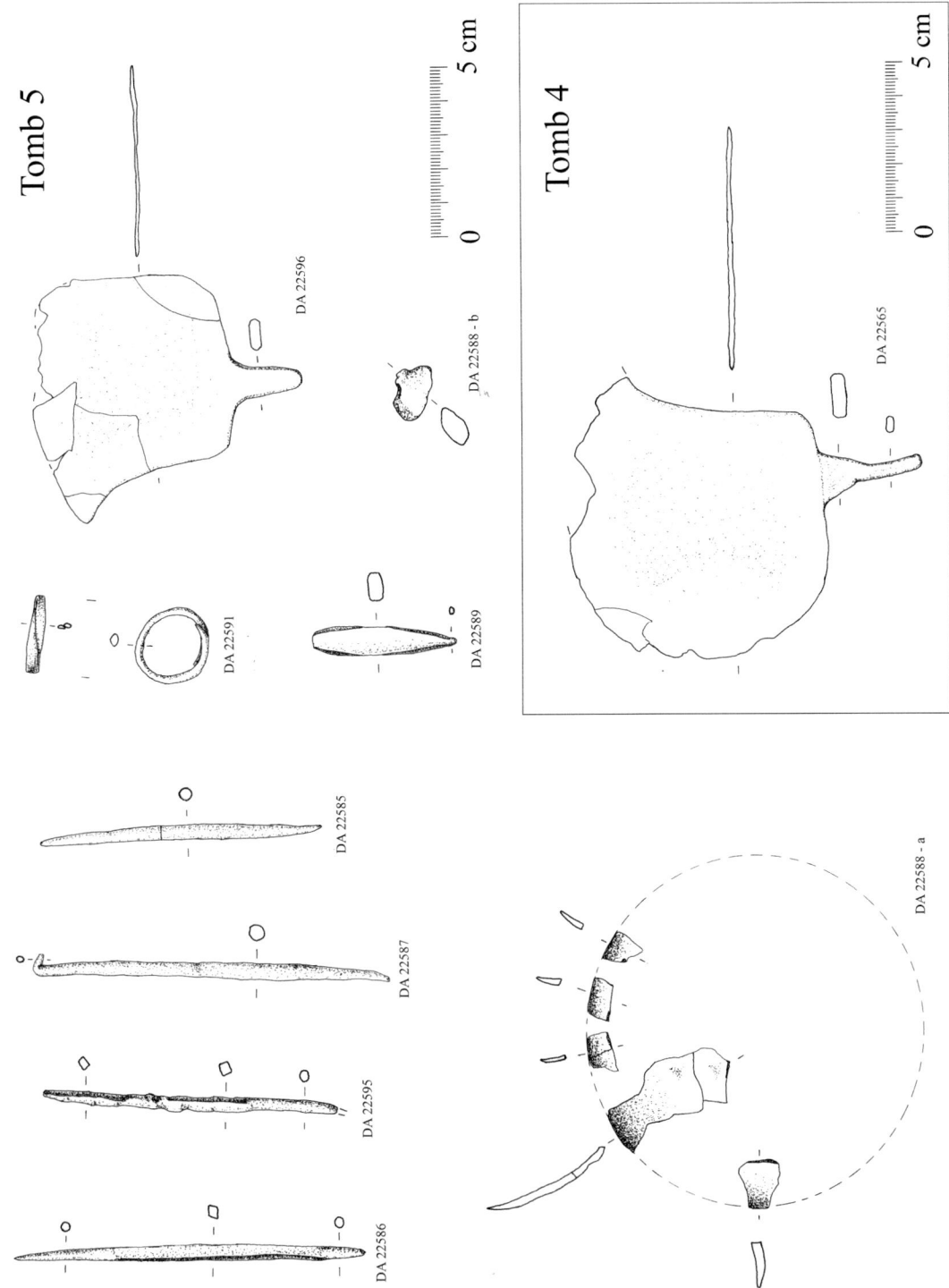

Figure 7.28. Ras Al-Hadd HD-7. Metal finds from tombs 4 and 5 (drawings by L. Tricarico).

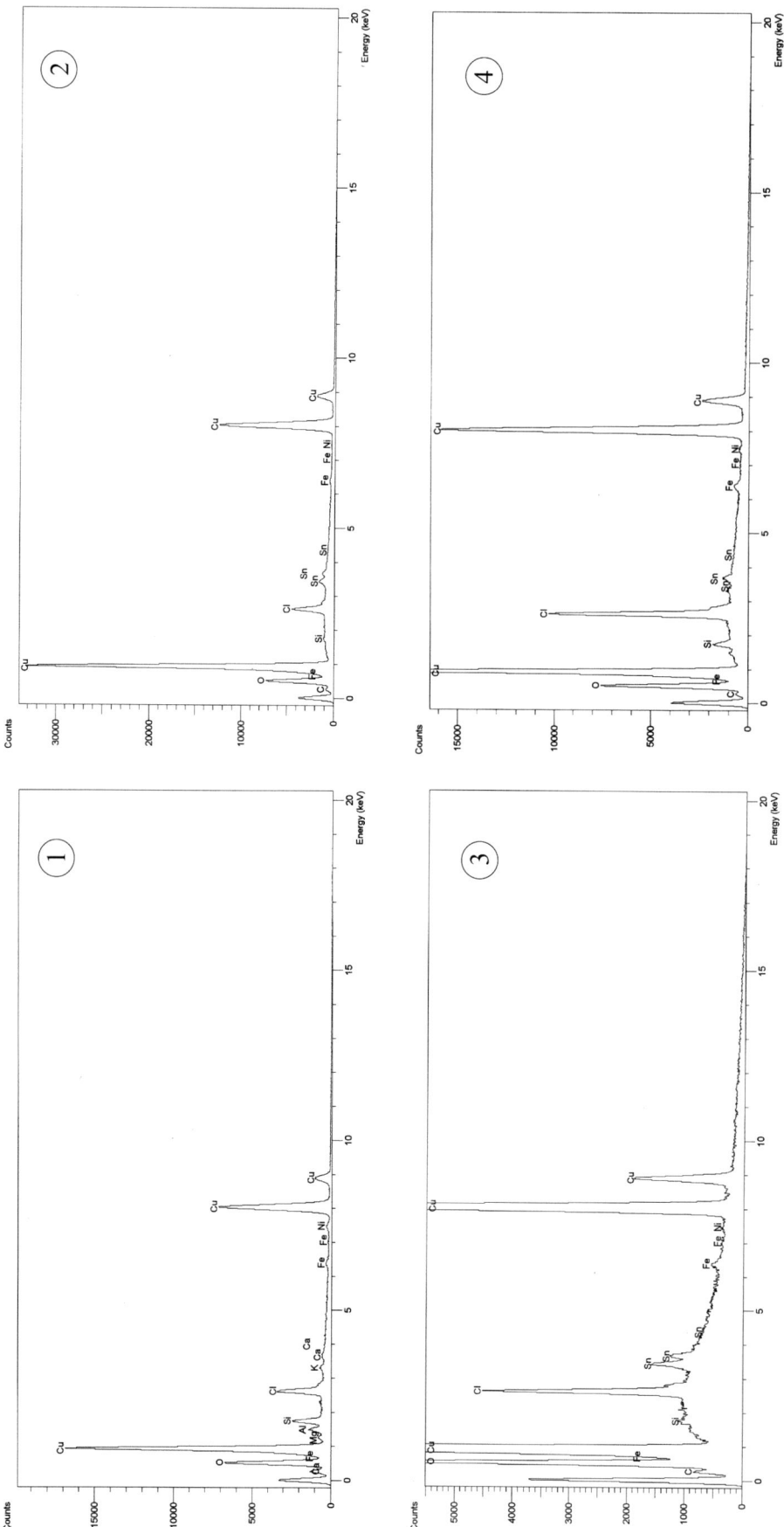

Figure 7.29. Ras Al-Jinz RJ-2. SEM/EDS spectra of the necklace from Building VII.

Figure 7.30. Ras Al-Hadd HD-5. Cockle shell containing a metallic cosmetic amalgam (photograph by C. Giardino).

Tin came to Oman thanks to long-distance trade. There is no unequivocal provenance for this metal, indispensable for bronze production. Geological evidence suggests that tin could come to the Arabian Peninsula from southern Anatolia, Central Asia and Afghanistan. To get to the Arabian Sea, Afghani tin had two possible overland routes: via the Baluchistan highlands or via southeastern Iran. Nevertheless, according to recent research, there are doubts that western Afghanistan tin sources were exploited before the end of the 3rd or the beginning of the 2nd millennium BC; other tin ore deposits are located in western Iran, Uzbekistan and Tajikistan (Thornton and Giardino 2012). Copper minerals were not exploited only for smelting activities in order to produce the metal: some of them were also traded as pigments, to be used as cosmetic powder, a kind of product highly appreciated in the Near Eastern urban centers.

A cockleshell valve from Ras Al-Hadd HD-5 – a settlement dated to the second part of the 3rd millennium BC – was found packed with the remains of a metallic, green, very finely ground amalgam, partially solidified. Most probably, it was a sort of *kohl* employed as an eye cosmetic (Figure 7.30). X-Ray Fluorescence analyses revealed that the amalgam was mainly composed of a mixture of copper sulfide and lime (CaO): it contained copper (79% Cu), with calcium (7.3% Ca), iron (7.2% Fe), chlorine (4.1% Cl) – probably connected with the sea environment – sulfur (1.1% S) and traces of nickel (0.2% Ni). The nickel content suggests that the ore came from one of the Omani deposits. Similar shells (*Anadara ehrenberghi*) used as cosmetic holders were also found at Ras Al-Jinz RJ-2, in a context of ca. 2300 BC. They contained a black or bluish-black amalgam of pyrolusite (MnO_2) and lime, which was used as a cosmetic coloring (Cleuziou and Tosi 2007: 175, fig. 181).

Chapter 8

The prehistoric copper mines

According to geological evidence, Oman is a country rich in copper ore deposits, mostly found in the Semail Ophiolite that characterize the mountain area of northern Oman. Remains of the old workings, mining dumps and slag heaps have been recorded by geologists and archaeologists since the late 1920s. Nevertheless, the real research for prehistoric mining started only in the Seventies thanks to the efforts of the German team from the Deutsches Bergbau Museum of Bochum led by Gert Weisgerber. He came first to Oman in 1977 taking part in Maurizio Tosi's Italian Expedition, to visit the mining and smelting sites of Arja, Al-Sayab, Al-Beyda, Lasail, Samdah and Samad 5. Weisgerber collected many slags, potsherds, fragments of tuyères and stone implements related to ore treatment. The mineral prospectors of Prospection (Oman) Ltd. had discovered 115 ancient smelting sites and had recorded the presence of large slag heaps dating back from prehistory to medieval times, even if most of them had an Early Islamic origin. Weisgerber promptly published a preliminary report of his surveys and started soon after a large project of archaeomining and archaeometallurgical research in the country (Weisgerber 1977; 1978).

From 1979 to 1982, the Deutsches Bergbau Museum of Bochum excavated the settlement of Al-Moyassar, where they found clear evidence of mining and smelting. This site had been discovered by the Harvard Archaeological Survey in 1973, which named it Samad 5 (Hastings *et al.* 1975: 12). According to the data from the German excavations, the metallurgical activity started at the site during the Early Bronze Age, in the 3rd millennium BC, as early as the Hafit period: it is therefore broadly contemporary of the fishing sites on the coast at Ras Al-Hadd HD-6 and Ras Al-Jinz RJ-2. The production reached its peak in the Umm an-Nar period. People from the site mined copper ore deposits located close to the settlement. There is in fact evidence of prehistoric open cast mines and pits in the slope of the hills located a few hundred meters from the site (Figures 8.1 and 8.2) (Weisgerber 1980 a: 78, fig. 28). Deep diggings of almost 10,000 m^3 testify that metal ore was quarried with an open casting technique. Other important mines were also discovered inside valleys to the west and south-west of Al-Moyassar. More than hundred Bronze Age graves were discovered around the settlement on top of hills and mountains. They probably belonged to the miners and metallurgists that lived at Al-Moyassar, controlled and protected by the ancestors from their tombs.

Prehistoric mining

The technique for the exploitation of metal ore deposits is linked directly to the previous Neolithic flint workings, where open-cast and underground mining were used to extract flint nodules. Despite many copper ore deposits easily available to the prehistoric miners, evidence of early exploitations is rare, because they were often destroyed or obliterated by later workings. Ancient copper mines had been discovered in many places of the Near East, such as at Timna in the southwestern Arabah (southern Israel), at Kozlu in Anatolia (northern Turkey) and at Ghar Chale in the Veshnoveh area (central-western Iran) (Rothemberg 1978: 1-2; de Jesus 1980: 109-110; Holzer *et al.* 1971: 4-5, 7).

Figure 8.1. Open cast mine at Al-Moyassar (Maysar) filled by mining debris (photograph by C. Giardino).

The prehistoric prospectors had much empirical knowledge, which enabled them to identify the metal ore outcrops. They were able to recognize the copper outcrops even at a distance because of the bright colors of the weathered rocks that characterize the upper and exposed part of the ore deposits, the so-called gossan, as in the Omani Ophiolites sulfides deposits. The peculiar heaviness of copper minerals and their vivid colors provided generally accurate advice on their metallic nature; more information could be obtained by smell and taste. Many copper ores, as chalcopyrite, release a characteristic smell when struck with a rock; the presence of arsenic in an ore is revealed by its peculiar garlic smell. Some minerals leave also a characteristic metallic taste on the tongue. Early mines were established just where the ore was outcropped. It was followed until the vein was lost or when digging became more difficult because of rock instability or water (Craddock 1995: 31-69). Prehistoric mining was carried out mainly with lithic tools such as stone hammers, which are precious indicators of early mines; they were intended to deliver a massive blow to the rocks.

Generally, their heads had ovoid shapes, and their weight was around 1 to 5 kg; they were often made with big cobbles of hard rock available in stream beds or *wadis*. Sometimes they had a continuous, large groove pecked around their midriff for hafting, while others had notches pecked into the midriff, but many stone hammers had no pecking at all (Figure 8.3). Frequently, these tools were found in mining dumps, as at Al-Moyassar. Experiments show that working directly with a stone hammer against a rock without a handle is extremely tiring and it damages the skin on the hands: therefore, most probably the natural shape of the cobble allowed it to be wedged in a haft even if it had no pecking (Pickin and Timberlake 1988). Concave and round strokes are the typical features left by hammer-stones on the walls of the mines: they give a relevant indication of prehistoric workings. After the spread of metal tools, in more recent times, these traces were replaced by long, deep and parallel grooves made by iron picks.

The mining tools were also made with wood and animal bones; they were preserved only in special micro-environmental conditions. Small shovels were made with large mammal scapulae and small chisels were obtained from long bones. Some examples of wooden shovels and hoes had been found in European prehistoric mines. The handles of many stone tools had to be made also with wood. Copper alloys played only a negligible role in the production of prehistoric mining instruments; some metal chisels were recovered in the prehistoric mine of Nujum near Bid Bid. For a long time, cheaper, traditional materials such as stones remained in use.

Figure 8.2. Copper workings at Al-Moyassar (Maysar). Collapsed entrance of a mine (photograph by C. Giardino).

Figure 8.3. Grooved stone-hammer from the mining area of Al-Moyassar (Maysar) (photograph by C. Giardino).

Ancient miners used wood fires in order to weaken the wall rocks and break them more easily: the so-called fire setting. Fires were set against the rock face to be worked and then left to burn for many hours (Figure 8.4). The intense heat produced fracturing of the rocks that could then be easily removed with primitive tools such as stone hammers and picks. Sometimes water could be used to extinguish the fire and to quickly cool down the shattered hot rocks, allowing the miners to start working. According to experiments carried out on fire setting techniques, the role of water was not necessarily to produce a thermal shock, because the exposed rock was already deeply fractured after the fire. The heat slowly penetrates into the wall and shatters the rock for up to 30 cm below the surface if the fire is left overnight (Craddock 1995: 33).

The result of fire setting was an irregular, smooth concavity in the rock; the workings are very characteristic, because the mine walls tend to have a rather sinuous aspect, following shallow curves. This technique has been observed in many prehistoric mines, both in Asia and in Europe. Mention of fire setting is also in the Bible (Jeremiah 23, 29); it still remained in use after the medieval period, until the 19th century. Fire setting is described and illustrated by Georgius Agricola in his treatise on mining and mineral extraction, De Re Metallica (English "On the Nature of Metals"), that was published at Basel in Switzerland in 1556 (Figure 8.5).

During prehistory, the metal ore deposits were exploited through surface mining (open-pit mines) or through underground mining (tunnels and shafts). The first technique was obviously easier because it did not require tunneling into the earth to follow the mineralized veins in depth. Torches or lamps provided lighting in the underground galleries. The main problems for underground mining were ventilation, the aquifers and the collapsing landslides. For ventilation, an easy system was to light a fire in the shaft in order to induce a draught thanks to the hot air rising; another one was to create a movement of air by moving linen cloths (Craddock 1995: 73-74) (Figure 8.6). Ancient miners carefully removed the extracted ore, often taking out all the mineralized rocks. Generally, the mined ore was first selected inside the mine; but the real selection occurred outside, near the excavation. The ore was coarsely crushed with stone-hammers, then ground with mortars and pestles; the gangue and the broken tools were thrown into the mining waste.

Early mines in Oman

Unfortunately, most of the early Omani mines were obliterated and destroyed by later miners, who repeatedly exploited the same copper ore deposits. A situation that occurs very frequently in the Near East and in Europe. Medieval mining flourished in the rich copper ore deposits such as Mullaq, Al-Lushal, Bilad Al-Maidin, Lasail, Arja, Semdah, Raki and Tawi Raki. In these places, the deeper layers of ore deposits had still a large amount of metal mineral left in Early Islamic times, after they had been already partially exploited during the 3rd and 2nd millennium BC.

Nevertheless, prehistoric mines have survived in northern Oman not only at Al-Moyassar, but also in other sites such as at Gebel Saleli, Huqain, Tawi Ubaylah near Nizwa, Wadi Miadin, Bilad Al-Maidin and Nujum near Bid Bid around Wadi Samail. Looking at this evidence we know that the local miners were expert in both surface and underground mining techniques. At Bilad Al-Maidin copper ore deposits occur inside the Ophiolites; they started to be mined during the Bronze Age, as testified by remains of an Umm an-Nar settlement and the coeval tombs close to the mines. Rock cuts, mining trenches, piles of broken ores testify the metallurgical activity on this site, operating from the 3rd millennium BC until the Iron Age and the Early Islamic period. Large slag heaps are an indication that also smelting operations were carried out on site (Weisgerber 2007a: 198-200) (Figure 8.7). The Bronze Age mines at Wadi Miadin followed the copper ore veins in depth; the prehistoric miners were only stopped by ground water. As it frequently happens in the early copper mines, the exploitation cavities were extremely narrow and not wider than needed.

A good example of an open cast mining is at Nujum near Bid Bid, where great trench mines are still preserved; these trenches were several hundred meters long. At Nujum some of the miners' tools were recovered, lying around these trench mines. They were mostly made of stones, but prehistoric workers used also metal chisels to break the rocks (Weisgerber 2007: 198, fig. 206; Weisgerber 2008). The stone-hammers carry evident signs of pounding at the end, which was the working face; some of the tools were discarded after their use when they were broken up into pieces. The examination of the stone-hammers from Nujum allowed already distinguishing different types, according to size and shape.

Figure 8.4. Experimental firesetting (photograph by C. Giardino).

Figure 8.5. Firesetting in Europe during 16th century (after Georgius Agricola, De Re Metallica Libri XII, Basel 1556).

The hammers were made out of stone cobbles most probably coming from the *wadi* bed. Some of them are big ovoid hammers, but there are also small pebble-like pieces, that were probably used until they broke into smaller fragments. Other tools have an almost spheroidal shape; others are thick, flat and pointed, and used as stone picks. Some of the very small pieces could have been employed as little pestles for small scale work, or else as *retoucher* to reshape bigger hammers or for the first ore treatment. The smaller hammers and the reused flakes would have been used for loosening the ore located in small crevices. Some have little notches at their sides, instead of the proper and deep grooved rills going all around the tools that are common features in the mining stone-hammers of other countries; these could probably have been hafted. Most of them have no hafting marks, but they show evidence of use at both ends; probably they had been wedged into a haft thanks to their elongated shape, even if it is also possible that they were hand-held. Some forms were used as wedges or chisels and only a portion seems to have been shafted with a handle. The typology of the hammer-stones from the mine of Nujum shows a strong correspondence with similar tools from Al-Hajir near Quriyat, southeast of the capital Muscat.

Figure 8.6. Bilad Al-Maidin. Rock cuts and broken ore fragments at a copper ore outcrop (photograph by C. Giardino).

Figure 8.7. Linen cloths used as ventilation system in Europe during the 16th century (after Georgius Agricola, De Re Metallica Libri XII, Basel 1556).

Flint and quartzite instruments were used here to quarry chert nodules from the limestone hill (Weisgerber 1997: 153-155, fig. 7). This is an evidence of the technological connection that links exploitation of copper and flint in prehistoric times.

Ore beneficiation

Before smelting, the ore had to be sorted in order to separate by hand the desirable mineral from the gangue, which is the useless part of the ore that has to be removed, concentrating the richer metal ore (Merkel 1985: 164-168; Giardino 2010: 55-56). These preliminary steps – if carried out accurately – made the smelting process easier and more profitable; they could reduce also the required amount of fuel and flux. They could also affect the chemical composition of the copper alloys, and therefore, ultimately, the quality of the final objects. A very initial sorting had to occur in the rough crushed mineral inside the mine by picking out the ore by hand.

Figure 8.8. Bilad Al-Maidin. Stone mortars and slag fragments scattered on the surface of the site (photograph by C. Giardino).

Generally, more precise crushing and sorting operations – the so-called beneficiation process – were carried out close to the mines. Large heaps made up of debris are found near the mines, with fragments of gangue of different sizes, frequently as peas or walnuts (Craddock 1995: 156-167). The copper ore was crushed with hammer-stones on heavy mortar stones; the surface of these tools has small circular depressions that were caused by pounding and crushing copper ore. This kind of tools was found together with mining instruments around the trench mine at Nujum. Small or large stone mortars with dimples on the face were found not only near prehistoric mines, but also at later sites. Sometimes mortars and pestles, associated with broken slag, are spread on large areas, as at Bilad Al-Maidin (Figure 8.8). The ore was again sorted by hands after crushing, evaluating the color and the weight. In many parts of the world, this work was traditionally performed by women and children.

Most of the pea size copper ore fragments were optimal to be smelted in furnaces. Sometimes – and mostly for precious metals as gold – the crushed pieces were ground again in order to turn them into a sort of powder: in this way, it was possible to separate the remains of the gangue from the metallic mineral by sieving it over water, washing away the lighter rock and leaving behind the denser and heavier metal.

Chapter 9

Copper smelting in prehistoric Oman

The metallurgical term *smelting* indicates the process by which a metal is obtained from its ore by heating it beyond the melting point. In antiquity, the ore was submitted to a preliminary beneficiation process before it was brought to the furnaces in order to turn it into metal; beneficiation consisted of removing by hand as much as possible of the gangue, i.e. the useless part of the ore.

The smelting process was carried out in a furnace, where it was possible to reach the reducing atmosphere and the melting point temperature (Figure 9.1). The melting point of an ore is generally lower than that of pure metal: for copper minerals, this temperature can also be around the 800 °C, although pure copper melts at 1083 °C (Giardino 2010: 56-57). Charcoal was the common fuel, which also provided the necessary, reducing atmosphere (Figure 9.2). Animal dung, from cows and camels, was not probably very significant in smelting operations, even if it is still used nowadays as a fuel in small earth smithies in many tropical countries (Craddock 1995: 196). Nevertheless, it should be considered that in arid or relatively arid climates – where wood may have been scarce – animal dung could have been a partial alternative source of fuel (Miller 1984).

Smelting is a complex chemical reaction that uses heat and a chemical reducing agent to decompose the ore, eliminating other elements as gasses or slag and leaving just the metal behind. Essentially, the chemical reaction can be summarized as follows:

Metal ore + Heat + Reducing agent → Metal + Gas + Slag

The reducing agent is commonly a source of carbon such as charcoal. The carbon (or the carbon monoxide derived from it) removes oxygen from the ore, which oxidizes it in two stages, producing first carbon monoxide (CO) and then carbon dioxide (CO_2) (reduction), leaving behind elemental metal. The absence (or at least scarcity) of oxygen is required for this chemical reaction to occur. This is obtained by surrounding and covering the ore with coal, so the mineral is located in the most reducing zone of the fire. In this way, the oxygen bound to the metal in the ore combines with the carbon monoxide given out by the coal: this is the way that metal and carbon dioxide are produced; CO_2, that is a gas, is blown away dispersed in the air.

Most ores are still impure, even after beneficiation, so it is often necessary to use a flux to remove the accompanying rock gangue as slag. Flux was constituted by iron oxides – as limonite, hematite or magnetite – in the frequent case of quartz gangues. If the original ore was made up from sulfides instead of oxides or carbonates, sometimes a preliminary treatment of roasting was carried out; the ore was piled on stacks of burning wood in the presence of air. It allowed the oxidation of the metal by removing not only sulfur, but also other volatile elements such as arsenic, antimony and bismuth. The obtained oxide was then submitted to the smelting process already described.

Figure 9.1. Experimental smelting furnace (photograph by C. Giardino).

Strong blasts of forced air inside the furnace were necessary to reach the high temperatures required by the smelting process. In the earliest stages of metallurgy, the air was supplied by blowpipes through which air was blown; they were tubes of cane or rolled leather with a nozzle of refractory clay at the end, as shown in reliefs and paintings of Egyptian tombs (Zwicker *et al.* 1985: figs. 2, 4-5, 8). At a later stage, bellows of flexible leather were in use, operated by hands or feet; they ended with a clay tuyère that was inserted inside the furnace wall to deliver the blast of air.

Archaeological evidence of ancient smelting processes is evanescent; it is rather easy to find slag heaps, the most noticeable indicator of a metallurgical process, but generally the furnaces left only weak traces in the soil and fragments of vitrified walls. The furnaces were manufactured only for the practical purpose of allowing metal smelting from the ore. Therefore, they were doomed to be dismantled or abandoned to decay when they stopped their function. Only rare evidence of their structures survived, because of the wall damage produced by the high temperatures and, sometimes, the need to recover the metal left inside the kilns.

Figure 9.2. Experimental smelting furnace loaded with charcoal (photograph by C. Giardino).

The earlier smelting furnaces were a little more than hearths bordered by stones, with a diameter of about 60 cm, where the smelting process was carried out with fairly primitive techniques, which did not provide slagging. However, in the Near East there is evidence of furnaces that were already using slagging techniques as early as the end of the 4th millennium BC. They were made of a concave pit about 40 cm deep with a half-meter stone superstructure built with 45 cm diameter above it. An example is the furnace discovered during the archaeological excavations at site 39 in the Timna Valley alongside the Arabah, ca. 20-30 km north of the Gulf of Aqabah, in the Red Sea.

These early metallurgists did not know the use of lining the furnaces, neither the tuyères nor slag tapping techniques, but they were aware of insulated smelting hearths, fluxing and ore dressing (Rothenberg 1978). Occasionally, the waste accumulated in the furnace necessitated interrupting the smelting operations in order to remove it; this caused the expenditure of fuel, preventing a large-scale metal processing.

Figure 9.3. Tapped slag fragment from Al-Moyassar-1 (DA 2219), with a smooth upper surface characterized by flow patterns (photograph by C. Giardino).

More advanced furnaces allow the easy removal of slag, which was tapped off – drained out of the furnace at liquid state through a specially constructed opening – and flowed out (tapping). Usually the fluid slag dropped off in a specific, large and shallow pit dug in front of the furnace to ease solidification, left there to cool before being hacked away and discarded. The tapped slags have a characteristic aspect that makes them easy to identify: they are heavy, black, with a smooth upper surface characterized by flow patterns, which are more pronounced when the fluid slag was highly viscous. As a rule, slag was tapped in a viscous state and it started to crystallize inside the furnace. The smelting slag from Al-Moyassar is tapped slag (Figure 9.3). It shows that developed furnaces were already in ordinary use during the 3rd millennium BC in Oman, an indirect, but unmistakable clue of massive copper production.

Bronze Age copper smelting in Oman

The earliest evidence of copper ore smelting in Oman dates back to the Hafit period, as early as the middle of the 3rd millennium BC. Some small slag heaps were found at Batin in the Wadi Nam near Ibra. They were recovered beside the modern road; remains of the settlement were just on the other side of the road. The slags were found together with mortar stones used for crushing the ore and the slag itself: in fact, some copper remained imbedded inside the slag with these early technologies, and therefore slag was commonly crushed in order to collect drops and prills of this precious metal. Thermoluminescence analysis carried out on the slag gave a date of about 2660 BC: Batin is therefore one of the older smelting places in Oman. Considering that about 10 tons of slag were recovered at the site, the produced copper could have been at least around one ton (Yule and Weisgerber 1996: 141; Weisgerber 2006: 194).

Large slag heaps found around the Omani Mountains show that more extensive metallurgical copper production started during the Umm an-Nar period and continued in the next Wadi Suq period. Slag heaps were recovered by the archaeological surveys at Al-Moyassar and in its area; there were even larger Bronze Age heaps, with more than 4000 tons of slag at Wadi Sahl, Assayab, Bilad Al-Maidin, Mullaq and Tawi Ubaylah (Hauptmann 1985: 34; Weisgerber 1980; Weisgerber 1983; Hauptmann and Weisgerber and Bachmann 1988: 35-36) (Figure 9.4). Frequently, dimpled stones were found scattered on the ground in association with slag: they are mortars and hammers whose purpose was to crush not only copper ore but also slag in order to recover the small copper prills contained within (Figure 9.5).

The black slag from Batin is similar to that from Al-Moyassar-1 dating to the Umm an-Nar period. This means that in Oman there was not a technical difference in smelting during the 3rd millennium BC. In the Near East and in Europe the earliest metallurgists started smelting copper with a long period of trial and error, beginning with crucible smelting in a rather primitive way. In Oman, apparently, this early metallurgical stage seems to have been bypassed.

Figure 9.4. Bilad Al-Maidin. Tapped slag on the surface of a slag heap (photograph by C. Giardino).

Figure 9.5. Bilad Al-Maidin. Stone mortars for crushing slag and ore (photograph by C. Giardino).

Metallurgists started making copper with an advanced method using furnaces that produced slag as a waste. It is possible that the very early stages of copper smelting – maybe dating to the late 4th millennium or to the very beginnings of 3rd millennium BC – have not yet been found; or else we should suppose that smelting techniques arrived in Oman from outside, in an already developed stage. Other evidence of Hafit slag comes from the earlier layers of Al-Moyassar and from Khashbah in the Wadi Samad, where it was recovered in a settlement together with flint tools (Weisgerber 2006: 194). Bilad Al-Maidin started its activity as a metal producing site in the 3rd millennium BC. Local copper ore deposits were mined since the Umm an-Nar period; large slag heaps prove that smelting processes were carried out in this place, from the Bronze Age to the Iron Age and the Early Islamic period (Weisgerber 2007a: 198-200).

Al-Moyassar (Maysar). An early metallurgical site

The best-known Bronze Age smelting site in Oman is Al-Moyassar (previously spelled Maysar), a site whose main activity was devoted to copper production. The copper production started in the Hafit period, but the main activity was carried out at a later phase, during the Umm an-Nar period. The Al-Moyassar Valley, in the southern part of the Al-Hajar Mountains in Central Oman, is located in an area characterized by outcrops of the copper rich Semail Ophiolite.

Figure 9.6. Al-Moyassar-1. Landscape (photograph by C. Giardino).

The archaeological survey recovered evidence of more than fifty prehistoric structures, both settlements and graves in the Al-Moyassar Valley, dating back from the Bronze Age to the Iron Age. Bronze Age evidence is situated in the south of the valley, while the Iron Age one is located on the north side, at the feet of the mountains. In this area, archaeological excavations carried out by a German team from the Bergbau Museum of Bochum that worked for three seasons, from 1979 to 1982. The site of Al-Moyassar-1 is located on the right bank of the central branch of Wadi Samah; its surface was spread with potsherds, stone tools, slag and furnace fragments (Weisgerber 1983) (Figure 9.6). The German excavation brought to light the remains of foundation walls of an Early Bronze Age tower made of mud-bricks with an outer wall made of stone, and smaller rectangular houses of stones and mud-bricks.

Figure 9.7. Photomicrograph of a copper prill (2 mm diameter) imbedded in a slag from Al-Moyassar-1 (photograph by C. Giardino).

The excavation had also uncovered a well shaft; the well reached the water table at about 13 m below the soil surface. About one hundred stone tools, such as hammers and grinding stones, together with a bronze needle were collected inside the well during the archaeological dig. Copper ingot fragments, fireplaces, charcoal, ashes, crucible and furnace pieces, molds, slag and a copper flat axe were recovered in the houses. The slags were carefully broken into small pieces in order to collect the copper drops and prills that were embedded in and under the slag (Figure 9.7).

The copper had to be re-melted in a crucible. Shallow, flat, circular holes in the ground were found beside the fireplaces: most probably they were the place where the fluid metal had to be poured from the crucible in order to make ingots. Generally, Bronze Age ingots were cast into these holes made in the soil; after cooling, they got their typical plano-convex shape, with a gas-blistered upper surface and rough lower surface. The furnace fragments from House 1 provided some detailed information about the smelting furnaces that were in use in Oman during the 3rd millennium BC (Weisgerber 1983: 270, 274).

The furnace was built inside a room, with a front hole for tapping the slag. The furnace was pear-shaped; it had a circular base with a diameter of about 40 cm and a reconstructed height of about 50 cm. Its interior could likely maintain a temperature around 1150 °C; the chimney let escape the gas produced by the smelting process. The furnace was filled with charcoal and ore fragments, mixed in several layers; during the long-standing smelting process it was refilled many times. Bellows provided air supply. The charcoal reduced the sulfide ore, as chalcopyrite, into matte, of which many fragments were collected during the excavation. Matte is a mixture of molten sulfides formed as an intermediate product during copper sulfide smelting. Instead of being smelted directly to metal, copper ores were usually smelted to matte – generally containing 40-45 percent copper along with iron and sulfur – which was then treated by a final reduction process to produce the metal. In this second stage inside the furnace, air is blown into the molten matte, oxidizing the sulfur to sulfur dioxide (SO_2) and the iron to oxide, which combines with a silica flux to form slag, leaving the copper in a metallic state. This process is called a matte process and it was used during the Bronze Age for treating the pyritic copper ore deposits in Oman, the United Emirates and Iran (Craddock et al. 2003: 107-108). The raw copper produced by this process normally contains several percent of sulfur together with iron and it has to be purified: it is called black copper (Craddock 1995: 149-150).

Oman had no large forests that could easily supply large amounts of wood for fuel production. Many archaeological charcoals were collected and analyzed during the Al-Moyassar excavations in order to determine the Bronze Age existing flora and the possible effects of copper exploitation on the woody vegetation (Eckstein *et al.* 1987). Archaeobotanists determined five taxa, Acacia, Prosopis, Ziziphus, Tamarix and Maerua – that are still part of the current tree and shrub vegetation of the area – plus Pistacia and Dalbergia that are missing today. According to their evaluation, the Omani annual slag produced during the Bronze Age (about 40 tons), required about 56 tons of charcoal or 560 tons of wood. This means that if the current growth conditions of Prosopis were the same as in the past, the demand for wood could be completely provided without problem.

Bronze Age copper ingots from Oman

In the Bronze Age copper was stored and traded as ingots; during the 3rd and early 2nd millennium BC the most common type of ingots was the plano-convex shape, although they could differ in size and weight. This form is spread in the Arabian Peninsula: similar copper ingots were recovered in Oman at Ras Al-Hadd, Al-Moyassar, Al-Aqir, in the United Arab Emirates on the Umm an-Nar Island, and in Bahrain at Qal'at Al-Bahrain, Saar and Nasariyah (Frifelt 1995; Højlund and Andersen 1994; Killick and Moon 2005; Lombard and Kervran 1989).

Similar ingots occur also in many other Near Eastern countries, in India and Pakistan, in continental Europe and in the Mediterranean. They can be made with raw copper, from smelting, or else by recycling scrap metal that sometimes could also be alloyed. The core of the plano-convex ingots is generally rich in gas blisters that allow breaking them quite easily by hammering. Their shape is connected with the easy technique for their production, pouring the liquid metal from the crucible in flat, round holes in the ground; reason for which the ingots are always different in precise form, size and weight.

One of the rare copper ingots hoards from Oman was recovered beside a fireplace in House 4 at Al-Moyassar. It consisted of twentytwo plano-convex ingots or their fragments; the hoard weighted more than 6 kg. The chemical analyses carried out on some of the items showed that they were made of almost pure copper (Cu content ranging from 89.60 to 96.99%); the main impurities were sulfur, iron, nickel and arsenic; except for sulfur, trace elements were on average below 1% (Hauptmann 1985: 80-83, tab. 21; Hauptmann *et al.* 1988: 41-42, tab. 4.1). Nickel content was rather high, ranging about 0.1-0.5%; arsenic too shows around the same levels as nickel. The presence of high sulfur (between 0.68% up to 4.14%) and impurities suggests that copper was not refined before casting the ingots: they were therefore made of raw copper.

Another hoard of similar ingots was discovered at Al-Aqir near Bahla among the foundation stones of a long prehistoric dam, forming a barrier for catching rainwater and preventing soil run-off. The whole area is marked by several Hafit and Umm an-Nar graves. Several metal items were recovered at the site.They were five anthropomorphic figures, a flat axe, a shaft-hole hoe, a cleaver and some ingot fragments (Weisgerber and Yule 2003). The Al-Aqir hoard consists of sixteen ingots whose weight ranges from 600 to 1980 g (Figure 9.8). Some of them (for example DA 7400) has visible remains of small artefacts on the outer surface, an indication that the copper came from recycling scrap metal. Most of them have an irregular protruding circular "ring" on the lower surface, below the rim. Analyses revealed that the core of those ingots consisted of slag instead of copper (Figure 9.9): a slag was therefore coated with copper and the characteristic "ring" was produced by this process. These "fake" copper ingots were easily identified by a skilled merchant: they could have probably been a votive, religious offering, connected with the construction of the dam.

Figure 9.8. Al-Aqir. Plano-convex ingot (DA 7406) (photograph by C. Giardino).

The ingots from Al-Moyassar, were made of almost pure copper (Cu content ranging from 94.6 to 98.8%), with traces of iron, cobalt, nickel and arsenic (Prange 2001: tab. 37). Nickel content ranged at about 0.05-0.5%; arsenic was around the same contents as nickel.

Another ingot comes from Ras Al-Hadd HD-1, on the coastal area of Oman, far away from the copper ore deposits of the mountains, 60 km west from Al-Moyassar (Craddock *et al.* 2003) (Figure 9.10). The plano-convex ingot from HD-1 (DA 11691) is about 8 cm in diameter and it weighs 590 g; it is dated back to the 3rd millennium BC, to the Early Bronze Age. Its composition (Cu 96%; As 2.09%; Ni 0.585%; S 0.37%; P 0.04%; Fe 0.033%; Co 0.014%; Ag 0.01%) shows that it has more arsenic and nickel than the items from Al-Moyassar, but is rather similar to many objects from the Hafit site of Ras Al-Hadd HD-6, which is slightly earlier than HD-1. The analyses carried out by metallographic microscope and Scanning Electron Microscope (SEM) confirmed the presence of large quantities of copper sulfide (Cu_2S), an indication that the ingot comes from the incomplete smelting of a copper sulfide ore. Nevertheless, the ingot has only a very low iron content, while the matte process produces a copper rich in both, sulfur and iron. In Oman, as in the mines near Al-Moyassar, there are some copper minerals with low iron content that could be exploited during the Bronze Age, as copper hydroxyl-sulphate and brochantite ($Cu_4[OH]_6SO_4$).

Figure 9.9. Al-Aqir. Section of plano-convex ingot filled with slag (DA 7407) (photograph by C. Giardino).

The product of the brochantite smelting would have been a mixture of copper and Cu_2S, similar to the metal of the ingot from HD-1 (Hauptmann 1985: 82; Craddock *et al.* 2003: 109-110).

An intriguing surprise came from the Lead Isotope Analysis carried out on six of the ingots from Al-Moyassar, as well as on four similar ingots from Al-Aqir. These analyses – at present the most reliable archaeometric examination to detect the original source of copper from archaeological items – show that their isotopic composition ($^{208}Pb/^{206}Pb$; $^{207}Pb/^{206}Pb$ and $^{204}Pb/^{206}Pb$; $^{207}Pb/^{206}Pb$) does not match the isotopic composition of Omani ores, suggesting that their metal did not derive from the copper ore deposits of Oman. There is no completely satisfactory explanation for this: it is possible that the copper ingots may have been imported from some other place. The presence of foreign copper in a region that is rich in copper ore deposits occurred also in other countries. For example, it happened in Sardinia in the western Mediterranean, where large concentrations of copper ores are scattered across the island and where there is evidence of a prehistoric exploitation of these metal resources (Giardino 1995: 140-148, 308-309). Nevertheless, a large number of oxhide ingots were imported from Cyprus during the Late Bronze Age, therefore Cypriot copper was used together with Sardinian copper for metal products (Gale 1991: 212-224; Muhly 2009: 26-30; Lo Schiavo 2009).

The lack of any source in the Near East with a similar isotopic composition is a problem. It is also possible that our sampling of Omani ores is not complete (Prange 2001: 91-98, 102, tab. 30; Weeks 2008: 91-94; Begemann *et al.* 2010: 153-154). Saudi Arabia has copper ore deposits with an isotopic composition similar to the Al-Moyassar ingots; but, there is no evidence of a prehistoric exploitation of that ore (Weeks 2008: 94).

Figure 9.10. Copper ingot from Ras Al-Hadd HD-1 (upper and lower sides) (photograph by C. Giardino).

Analyses of slag

Smelting slags are made by iron silicates; therefore, they are mostly rich in iron and poor in other metals, as copper. Because they are the waste material of the pyrotechnological activity, their analyses provide a large amount of information on ancient smelting technology; therefore, they are a basic source for these data, also considering that furnaces are generally lost. Slag study is a powerful tool to investigate the condition that occurred in the prehistoric metallurgical processes. Hafit and Umm an-Nar slags from Al-Moyassar are similar for bulk chemical composition and microstructure (Hauptmann *et al.* 1988: 37-40). Metallographic and SEM analyses show that their main component is fayalite-rich olivine, whose small crystals are imbedded in a glassy matrix; microscopic copper prills were imbedded in the slag (Figure 9.11).

They have a similar content in SiO_2, iron oxide, MgO, CaO. Considering the siliceous gangue of the local copper ore, that content shows that iron minerals were used as flux, as limonite and hematite. The main difference between earlier and later slag is in their copper content: Hafit slag contains up to 31% of copper, while the Umm an-Nar slag generally has less than 2% of copper, reaching sometimes 7% Cu. This means that smelting technology improved enormously in the span of time between the two periods, becoming more and more efficient.

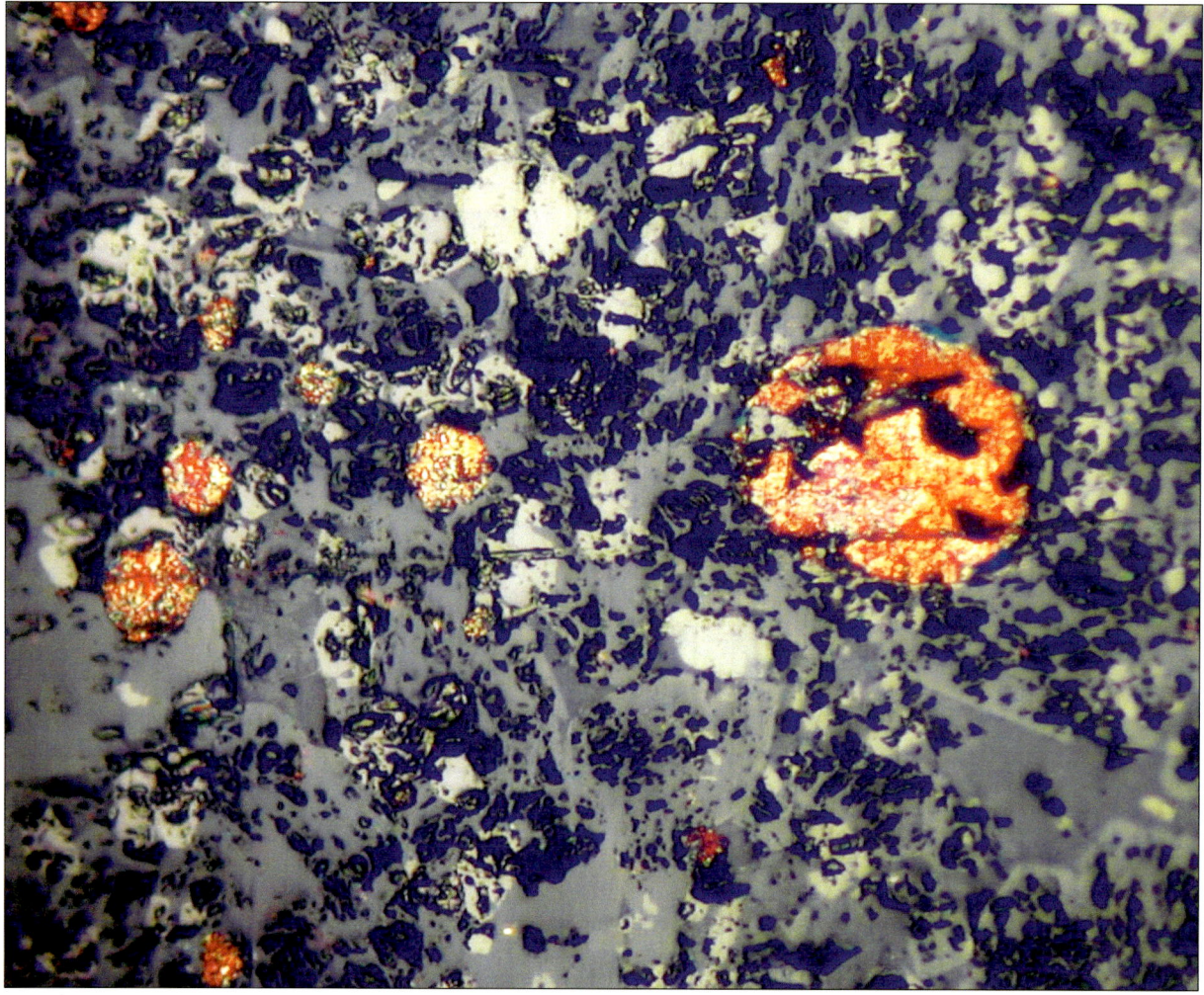

Figure 9.11. Al-Moyassar-1. Photomicrograph of copper slag (DA 2219) containing copper prills (20 x) (photograph by C. Giardino).

The analyses of the micro-texture of the iron-rich slag confirm this technical improvement: the earlier slag crystallized in rather high oxidizing conditions, while most of the later ones were produced in a strong reducing atmosphere. The chemical composition of the slag allows to estimate the firing temperature inside the furnace, looking at the temperature requested by some of the component to be melted, i.e. to change their physical status from solid to liquid. The studies on the slag from Al-Moyassar indicate a minimum temperature around 1100 °C as the possible working temperature of the furnaces. This data is confirmed by the analyses carried out on fragments of furnace lining from the site, which show that the smelting structure worked for several hours at a temperature around 1150-1200 °C. Anyway, it seems that these high temperatures could be reached and maintained for a short time only.

Chapter 10

Middle and Late Bronze Age: the Wadi Suq period, ca. 2000-1300 BC

The well-developed society established in Oman in the 3rd millennium BC reached its peak in the second part of the millennium, when the country entered written history thanks to the trade between Mesopotamia and Magan. Copper was one of the main materials of these trades; they introduced Oman in a larger system that connected Arabia with Syria, Iran, Mesopotamia and the Indus Valley (Cleuziou and Tosi 2007: 213-215). This complex socio-economic structure collapsed around 2000 BC, with dramatic changes. The presence of rich collective tombs is an indication of a complex and elaborate social organization during the Late Bronze Age. Nevertheless, the economic bases of the Omani societies appear much less connected with international trade than during the previous periods (Yule and Weisgerber 2016: 43).

In the first centuries of the 2nd millennium BC, the Mesopotamian metal trade with Oman ended, substituted by Cyprus as the main source of copper for the Near East. Copper ore deposits are located, in Cyprus as in Oman, in the Ophiolites. The geology of Cyprus is dominated by the Troodos Ophiolite Complex, a massif that covers about one third of the island. Massive cupriferous sulfide deposits were formed inside the layers of the pillow lavas. These ores consist almost of pyrite and chalcopyrite (Constantinou 1982: 13-17). The exploitation of local copper sulfides dates back to at least the Middle Cypriot, ca. 2000-1600 BC (for Cypriot chronology, see Rapp and Swiny 2003: 7), according to the evidence of mining at Ambelikou-Aletri and smelting at Pyrgos-Mavrorachi (Merrillees 1984; Giardino 2000) (Figures 10.1 and 10.2). Intensification of copper production occurred in 1700-1400 BC at the production centers of Enkomi – a dominant site in the developing copper industry – Kalopsidha Koufos and probably Hala Sultan Tekke, while copper sulfides were smelted at Politiko Phorades in the 16th century BC (Knapp 2013: 406-408). During the Late Bronze Age, Cyprus had a pivotal role for metal trade in the Mediterranean and in the Near East. The Amarna letters are a diplomatic document dating to the 14th century BC; they testify to the exchange between Pharaonic Egypt and the rulers of Alashiya, Cyprus, to supply Egypt with copper. Alashiya sent to that country on one occasion almost 3000 kg of copper (Muhly 2005: 139). The shipwreck of Uluburun in southern Turkey provides good information about Mediterranean metal trade in the Late Bronze Age; the ship sunk ca. 1300 BC. The ship cargo consisted mostly of raw material: 10 tons of copper ingots in the "oxhide" shape characteristic of the Mediterranean and one ton of tin ingots. According to archaeometric studies, copper ingots were manufactured with Cypriot metal (Pulak 2000).

During the 2nd millennium BC, in Southeastern Arabia, the new cultural aspect that characterizes the Late Bronze Age is called the Wadi Suq culture and it appears to be totally different from the previous Umm an-Nar civilization. The name comes from a graveyard of single burials located west of Sohar, where there cultural traits were identified for the first time. These tombs contained new types of metal objects and pottery that date to the beginnings of the 2nd millennium BC. The sites of the Wadi Suq period, both settlements and burials, are spread all over Oman, from Musandam to the island of Masirah, even if their distribution shows that sites are rarefied in comparison with the high density of the previous Bronze Age periods (Cleuziou and Tosi 2007: 257-278, fig. 274; Yule et al. 1994). Single and collective burials characterize the Wadi Suq funerary rituals and architecture.

Figure 10.1. Cyprus, Mitsero (Nicosia). Ancient copper mines (photograph by C. Giardino).

Figure 10.2. Cyprus, Mitsero (Nicosia). Ancient tapped copper smelting slag (photograph by C. Giardino).

In both cases, the tombs contained grave goods that frequently included metal weapons as daggers, spearheads, rapiers. Single burials were subterranean, in stone cists, in order to create a chamber covered by flat slabs and surrounded by a tumulus. Tombs of this type were excavated at Wadi Sunaysl (Ibri), Baushar, Khudra and Al-Moyassar in Wadi Samad, Sachrut Al-Hadri (Masirah). Collective burial had a different kind of architecture. The main type was a long, narrow rectangular chamber, with a single entrance on a long side, without the division into two non-communicating parts that characterized the earlier funerary monuments. Wadi Suq collective burials were found at Qattarah near Buraimi, Al-Qusais (Dubai), Al-Wasit, Shimal, Dhayah, and Ghalilah.

The collective tomb W1 from Al-Wasit is of special interest in the studies on metal artefacts of the Wadi Suq period. It had a rich inventory and it could be regarded as one of the greatest complexes of metal objects for this period. Tomb W1 contained mostly stone vessels and weapons (daggers, swords and spearheads) together with razors, silver bangles and rings. According to the analyses carried out on 49 copper-based objects from this tomb, they were mostly made of pure copper, with little alloying; arsenic is generally low, below 1%, and it was most probably a natural component of the copper ore. Only four artefacts contained tin, with a content ranging from 3 to 9%, an indication that tin was rare and precious in Oman until the Iron Age (Yule and Weisgerber 2015: 14-20, 43, tab. 22, pls. 7-41).

The settlement of Ras Al-Jinz RJ-1 is an important source of information on Wadi Suq metallurgy. The site is located on the central mesa of Ras Al-Jinz, in the Ja'alan coastal region, and it was occupied after the abandonment of RJ-2 (Figure 10.3). The remains of some thirty houses were recovered at RJ-1, which formed a dense cluster; they had one or two rectangular rooms (Figure 10.4). The village was protected from eventual enemies by the high, inaccessible slopes of the mesa, and a wall fortified the only access to the site. The economic system of RJ-1 was based on the exploitation of sea resources, catching and processing fish, exactly like the previous village built at RJ-2.

Types of artefacts from Ras Al-Jinz RJ-1

The few, small metal items that were found at Ras Al-Jinz RJ-1 reflect traditional fishing activities, and therefore they do not differ substantially from the objects that characterized RJ-2 (Figure 10.5). There were fishhooks, together with awls and needles for net making. The production of small metal tools that fishermen needed for their work occurred in the village. There is evidence of copper melting at the site, as testified by small casting residues found *in situ* (Figure 10.6). Some of the finds were probably obtained by recycling older, broken objects by hammering and sharpening. None of the few analyzed items had been produced with tin bronze: they were all made of almost pure copper, sometimes with high nickel content, up to 2%. A socketed spearhead belonging to a Wadi Suq type was found on site, but in a secondary position, ca. 2 km from the site (Cleuziou and Tosi 2007: 265).

Tools

> *Fishhooks.* The studied fishhook has strait shanks and it does not differ from similar tools from RJ-2. It was produced with a nickel-copper alloy (2.1% Ni) and only traces of arsenic (0.3% As) [DA 14406].

Middle and Late Bronze Age: the Wadi Suq period, ca. 2000-1300 BC

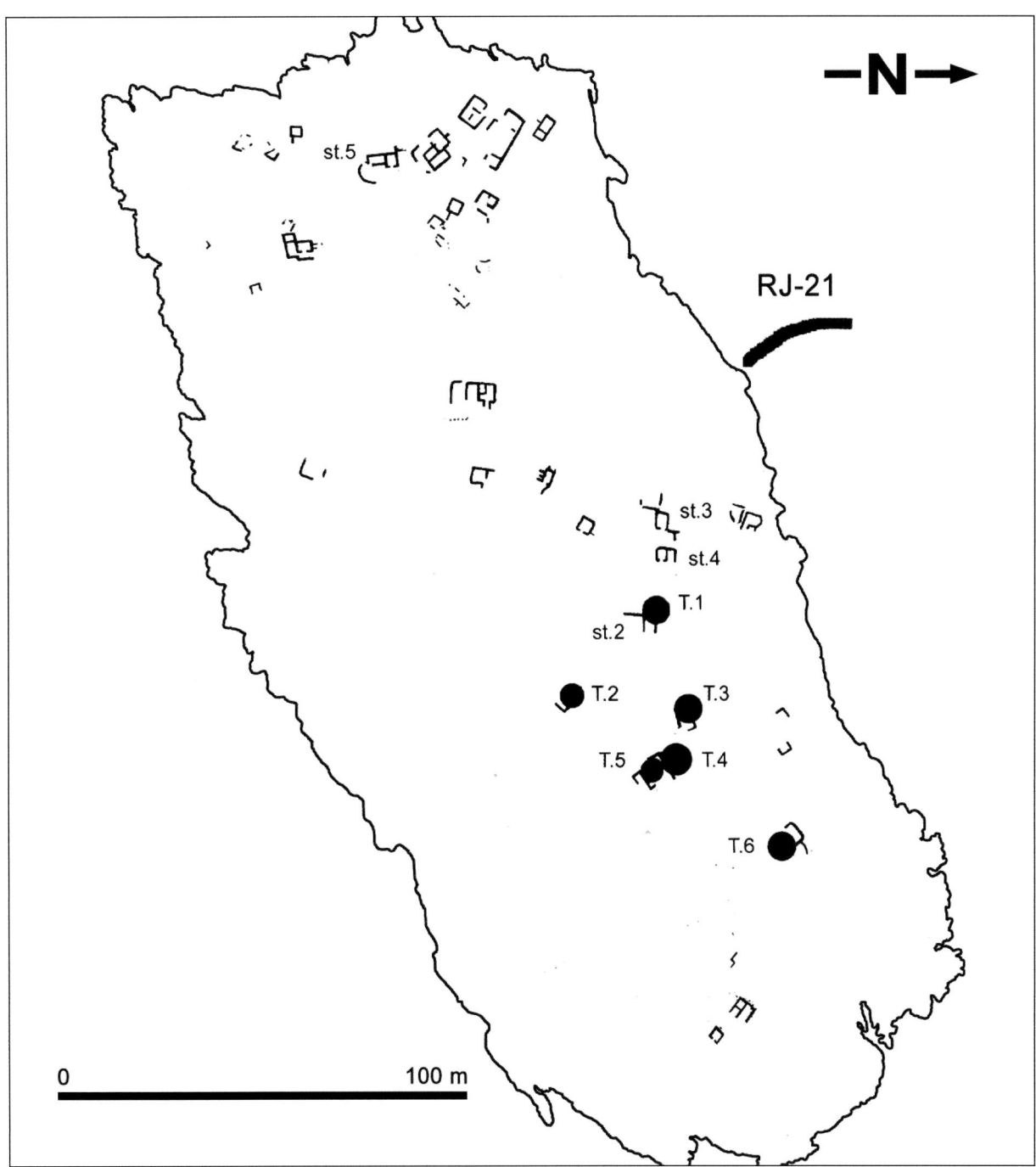

Figure 10.3. Plan of Ras Al-Jinz RJ-1 (after Cleuziou and Tosi 2007).

Figure 10.4. Ras Al-Jinz RJ-1. Wadi Suq house during excavation 1 (after Cleuziou and Tosi 2007).

Awls. The awl is rectangular in section and bent. It was made with pure copper (99.6% Cu) [DA 8681].

Needles. The needle from RJ-1 has an eye. It was produced with a nickel-arsenic copper alloy: 2.2% Ni and 1.4% As [DA 14313].

Flat chisels. Made of arsenical copper with only 2% As [DA 14404].

Ornaments

Spiral beads. Long bead made of a spring wire. Analyses detected traces of lead in almost pure copper [DA 14314].

Semi-finished products and fragments for recycling

Casting residues. The copper piece, irregular in shape, contains a relatively high content of iron (1.4% Fe) [DA 14405].

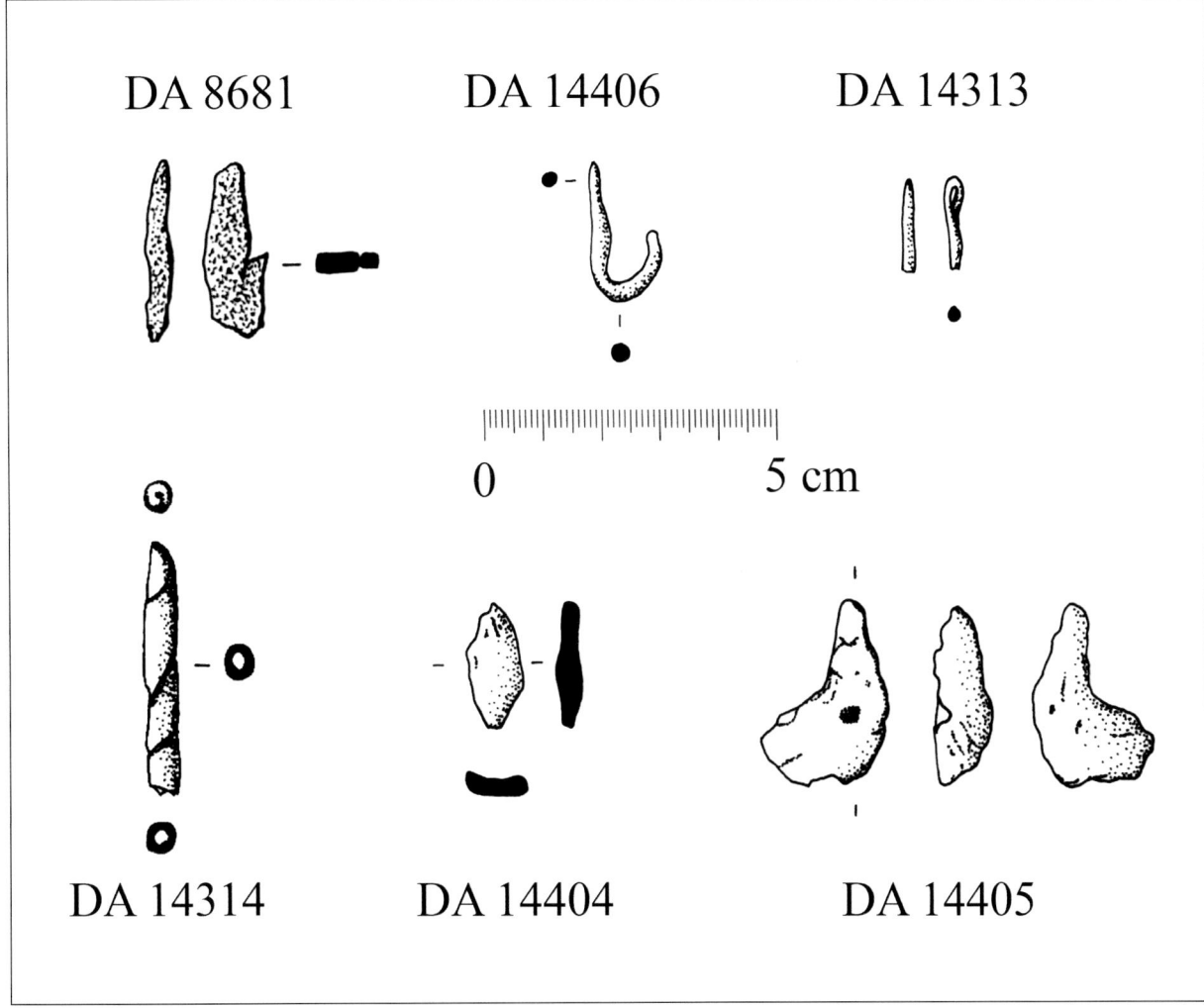

Figure 10.5. Metal objects from Ras Al-Jinz RJ-1.

Copper production in the Wadi Suq period

Copper production continued into the Late Bronze Age, using a similar smelting technique already developed in the past centuries. At Bir Kalher near Rawdah (Wadi Samad), about 20 km north of Al-Moyassar, a small slagheap was investigated by the German team of the Bergbau Museum (Kroll 1981; Weisgerber 2007a: 195). Slags were about 200 tons and many fragments of furnace were collected together with slag. These finds belonged to smelting operations that occurred in the Wadi Suq period, one very rare evidence of copper production from that time. The furnace fragments recovered at Bir Kalher are almost similar to those found at Al-Moyassar-1. The only difference seems to be in the external aspect, because the Late Bronze Age furnaces had a sort of foot at their bottom. Anyhow, this small variance did not interfere at all with the smelting processes. In fact, the slag produced at this site did not differ from the slag from Al-Moyassar, an indication that the smelting process was the same in the two places, even if the finds from Al-Moyassar are much older than those from Bir Kalher. It is possible that also the slag found at Wadi Sahl in Wadi Samad belongs to the same period; at the site, subterranean cist graves were found near the slag, beside copper ore outcrops.

Wadi Suq communities had an easy access to metal resources, as indicated by the richness and wealth of the graves furniture in some of the tombs. At Al-Wasit in the Wadi Jizzi area, the rich collective grave W1 contained 42 spearheads and 16 swords and daggers; two of the swords were 60 cm long (Bakhit Al-Shanfari and Weisgerber 1989: 18, pl. 5; Weisgerber 2007a: 195-196) (Figure 10.7). Not far away from the tomb an ancient slag dump was recovered, connected with the exploitation of the local copper ore deposits. Analytical studies were carried out on the metal finds from the Al-Wasit tomb; these data give new light on the economy and trade of Oman during the Late Bronze Age. Most of the 51 metal weapons from the grave furniture were analyzed by ICP-OES (Prange 2001: tabs. 32, 36). Looking at the composition results, the largest majority of the objects was made of almost pure copper, some with a nickel content up to 3%.

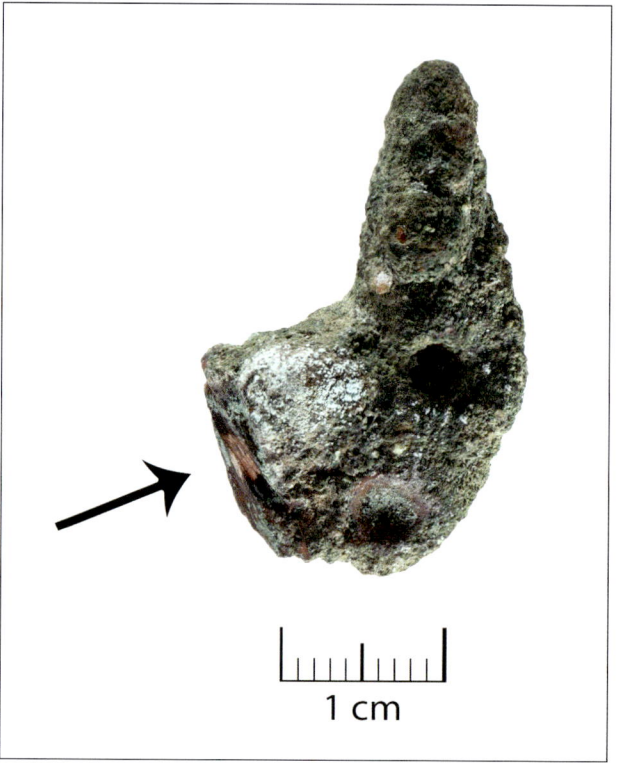

Figure 10.6. Casting residue from Ras Al-Jinz RJ-1. The arrow shows the cleaned point for the XRF analysis.

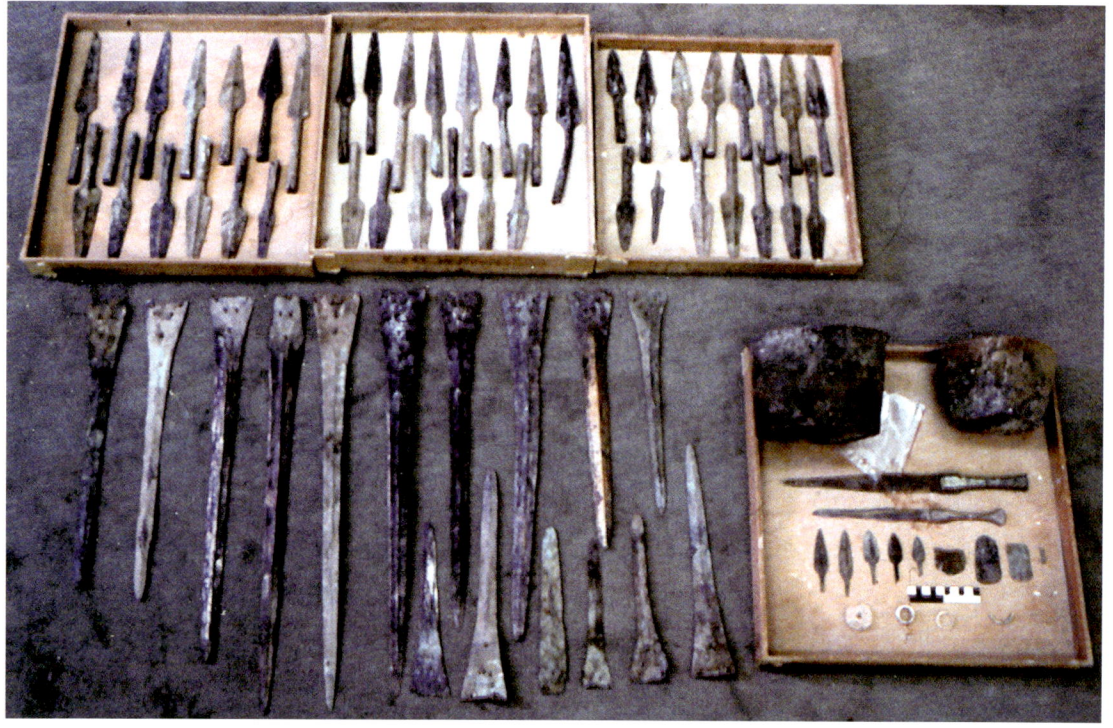

Figure 10.7. Wadi Suq metal objects from the Al-Wasit grave (after Weisgerber 2007b).

Only four pieces were made of tin bronze, i.e. less than 8% of the all items: two daggers (DA 07848 and 07849) and two spearheads (DA 07807, 08194) that had, respectively, 9.48%, 8.21% and 7.59%, 3.18% Sn. It means that access to the copper resources was easy, because this metal was locally produced, but tin and its alloys were really unusual. Bronze was used for manufacturing only some of the spearheads and daggers, but not for the swords. Spears generally suffer less stress during fighting than swords or daggers, so they do not need a very tough alloy. It is possible that spears and daggers were made in bronze only because they both were relatively small, so it was necessary to produce them with a smaller amount of tin or bronze compared to bigger weapons. Therefore, probably the alloy was not employed for its mechanical properties – tin lowers the melting point of copper, increases the fluidity of casting and improves mechanical properties such as malleability and toughness – but to "improve" the final color, in order to obtain a metallic golden-brown hue considered more suitable for precious, prestigious items devoted to emphasize social status.

A similar situation occurred also in more ancient times: bronze was reserved for ornaments in the Umm an-Nar period too. Tin was rare and precious, also for members of the elites that were buried inside the Al-Wasit grave. This is an important clue that Oman was not fully included in the circuits of international trade in which tin traveled towards the Near East and the Mediterranean during the Wadi Suq period.

Chapter 11

Early Iron Age, ca. 1300-300 BC

After the sunset of the Egyptian and Hittite empires that had controlled the eastern Mediterranean for much of the 2nd millennium BC, the ancient Near East went through a period of radical transformations during the centuries between 1200 and 800 BC. These transformations prepared the complex geopolitical systems that characterized the 1st millennium BC in the Near East. At the end of the Wadi Suq period, ca. 1300 BC, new transformations began in Oman, that culminated around 1000 BC in the creation of a new regional culture. Metallurgy – especially bronze metallurgy – played a fundamental role in the definition of the Omani Iron Age culture. This long-time horizon is conventionally divided into three periods: Early Iron Age I (1300-1100 BC), Early Iron Age II (1100-600 BC) and Early Iron Age III (600-300 BC). The second one is also known archaeologically as the Lizq period.

Indeed, the Iron Age in Oman still shows a remarkable level of uncertainty. This may be due to a notable level of ambiguity because of regional characteristics and the questionable dating of several sites. Thus, a subdivision in phases would be largely artificial. Profound societal changes only occurred around 300 BC, after the fall of the Achaemenid Empire at the hand of Alexander the Great (Cleuzious and Tosi 2007: 281-298; Yule and Weisgerber 2015: 42-43).

Early Iron Age I, ca. 1300-1100 BC

Until a few years ago, the most remarkable metal assemblage was a group of metal objects discovered in the late Seventies in a single grave on the slopes of a limestone mountain, the Jabal Al-Hawrah, east of Nizwa, above a *wadi*. The burial chamber was located between two rocks. Inside the tomb were found a battle-axe/halberd, three daggers, 27 arrowheads, a razor, a finger ring, and two copper alloy bangles, together with three softstone vessels, a pottery bowl, a calcite stamp seal, and a few carnelian beads (Figure 11.1). The tomb was interpreted as the grave of a wealthy, socially prestigious warrior (Bakhit Al-Shanfari and Weisgerber 1989; Cleuziou and Tosi 2007: 281-283, 292-292, figs. 299, 313).

The arrowheads started to appear during the 2nd millennium BC and became a rather common item inside tombs when weapons were buried there. For instance, a large collection of arrowheads was also recovered at Daba, in the Musandam Peninsula, where very rich Early Iron Age collective graves, almost contemporaneous with the tomb at Nizwa, were recently found. The arrowheads from Nizwa were not typologically homogenous; nevertheless, they may represent the content of a quiver. The widespread presence of arrowheads as grave furniture underlines the military importance and the high role of archers in the Iron Age Arabian society. The axe/halberd from Nizwa has parallels at Daba and shows analogies with similar objects from Luristan. According to compositional analyses – two of the daggers and the axe/halberd were analyzed by ICP-OES – the weapons were now made with tin bronze: the daggers DA 7720 and DA 7784 had, respectively, 10.0% and 7.86% tin, while the axe/halberd DA 7719 had 8.57% Sn (Prange 2001: tabs. 32, 36).

Early Iron Age, ca. 1300-300 BC

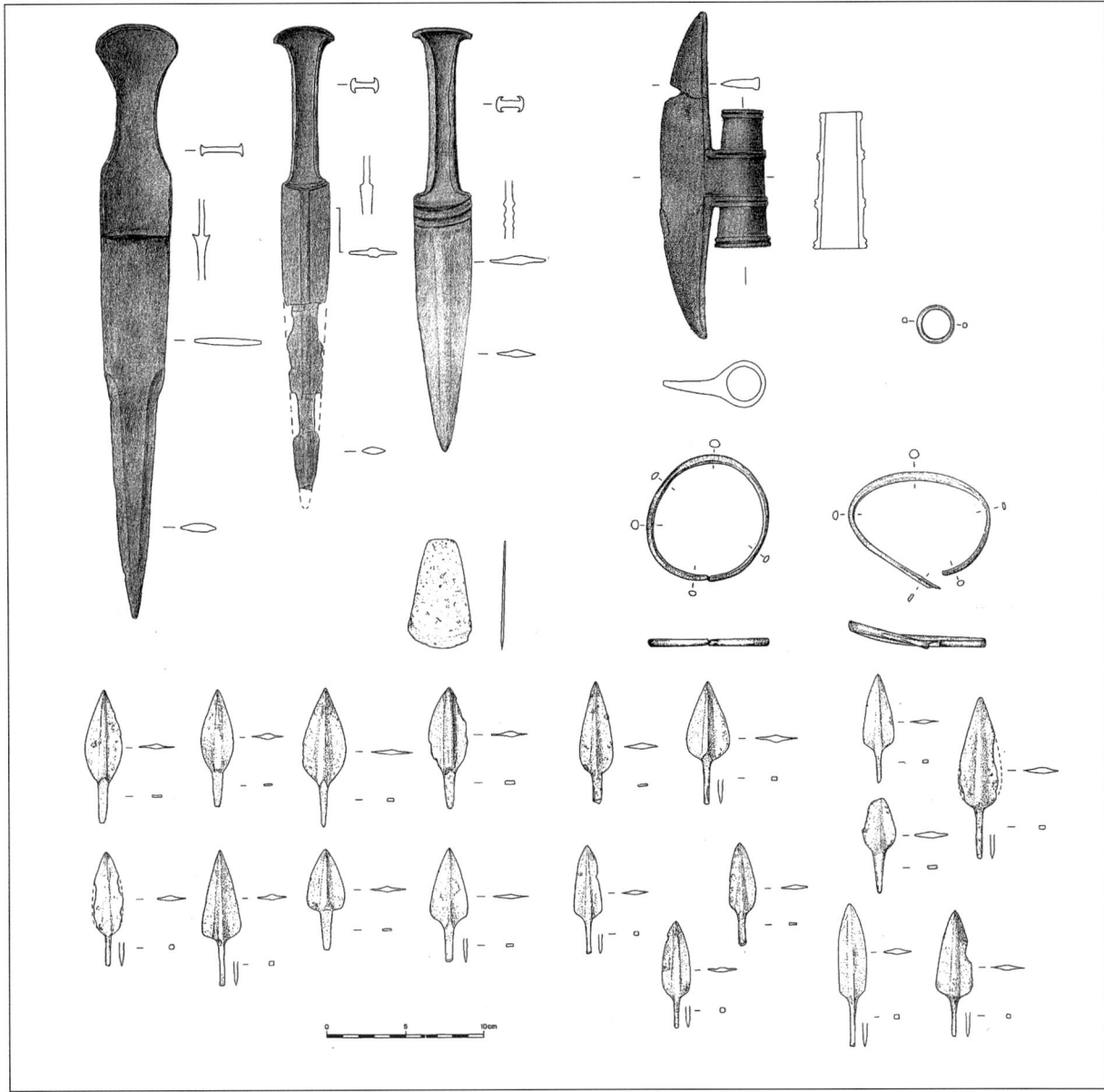

Figure 11.1. Metal objects from the tomb of Nizwa (after Al-Shanfari and Weisgerber 1989).

Iron Age metallurgical sites

During the Iron Age slag occurs in large cakes weighing around 10 kg, instead of the smaller slag fragments that characterized the Bronze Age sites (Weisgerber 2007a: 196-200). It is possible that a technological difference was connected with the adoption, in the Arabian Peninsula, of more developed multi-step chalcopyrite smelting techniques developed in Cyprus.

A large smelting place was recovered at Raki 2 near Yanqul, whose activity can be dated between 1300 BC and 800 BC. The site is close to the mining areas of Raki 1 and Tawi Raki, where there are remains of ancient mines. The evidence from Raki 2 is very impressive: about 10,000 tons of Iron Age slag covered the area; houses were built on the slag heaps using large slag cakes to raise the walls.

Radiocarbon dates from the excavation of the building complex found at Raki 2 on the bank of Wadi Raki gave dates from 1000 up to 840 BC (cal 2 σ) (Weisgerber and Yule 1999: 115). Fragments of Iron Age furnaces were collected at a small smelting site in the Wadi Qatof, west of Al-Moyassar: they belonged to almost cylindrical structures, 50 cm in diameter. Metallurgical activity was recovered also at Arja (Qarn Al-Muallaq), Wadi Miadin near Hawqayn (Rustaq), Jebel Saleli (Musfa) and Bilad Al-Maidin.

A relevant smelting site was recovered at Saruq Al-Hadid in the Emirate of Dubai; the place is near the border between Dubai and Abu Dhabi and it is characterized by sand dunes (Qandil 2005: 126-138). At present the lack of water resources makes the area uninhabitable; the rare vegetation is composed of a few trees of *Prosopis cineraria* and some bushes. Saruq Al-Hadid is characterized by an intensive scatter of archaeological material on surface, including large amounts of smelting slag, copper ore fragments, crucibles, basalt tools such as grinding stones, hammers and mortars. The excavation allowed collecting potsherds and stone vessel fragments datable to the 1st millennium BC, together with a large number of copper-based objects such as metal bowls, axes and arrowheads. Saruq Al-Hadid was a lucky opportunity to analyze a well-dated metallurgical site. Generally, it is hard to chronologically distinguish the different smelting products of the Iron Age without the help of other archaeological finds, as pottery or metal objects.

Two new discoveries: 'Uqdat Al-Bakrah and Daba

Recent excavations carried out in Oman by the Ministry of Heritage and Culture, at Daba and at 'Uqdat Al-Bakrah, have given new light about the socio-cultural development and metallurgical production in the country during the Early Iron Age. Before this new evidence, the knowledge about the first period of the Omani Iron Age was relatively scarce. The archaeological sites of 'Uqdat Al-Bakrah (previously known as As-Safah or Al-Ṣafā') and Daba (or Dibba) were discovered by chance in 2012: they both opened new perspectives on the complex political, social and economic aspects of northern Oman in the Early Iron Age. Both 'Uqdat Al-Bakrah and Daba inform us about a tribal society of great material opulence that dates back between 1110 and 800 BC. Thousands of metal items, including precious metals, carved stones and more are sacrificed to the afterlife in large collective graves of Daba, buried in the ground with a huge waste of goods. The large collective grave LCG-1 of Daba contained ca. 4000 objects; half of them were luxury items, including silver and a gold piece of jewelry. 'Uqdat Al-Bakrah is a metallurgical workshop site discovered in the Rub Al-Khali desert.

'Uqdat Al-Bakrah: an Iron Age metallurgical workshop

The evidence from 'Uqdat Al-Bakrah give us relevant information about a metallurgical activity during the Early Iron Age. At 'Uqdat Al-Bakrah were in fact discovered the remains of a metal workshop that was mainly devoted to casting activities. The site was discovered in 2012. It is located on the first dunes on the northeastern edge of the Rub Al-Khali, about 40 km from the oasis of Dank (Figure 11.2). The archaeological excavation distinguished two main areas, called 'Uqdat Al-Bakrah 1 and 'Uqdat Al-Bakrah 2, separated by a dune belt. The distribution of objects was concentrated around clusters of metallurgical structures at 'Uqdat Al-Bakrah 1, while 'Uqdat Al-Bakrah 2 had a very high concentration of different types of furnaces.

Figure 11.2. The site of 'Uqdat Al-Bakrah in the sands of Rub Al-Khali (photograph by P. Koch).

Hundreds of objects, mostly tools, weapons and metal vessels, are one of the most intriguing archaeological discovery of this century. All the items were destined to be re-cast, because the workshop was devoted to the production of new objects, like axes, by recycling old metal items. The location is among the dunes in an unwelcoming place far away from the oasis and the main settlements. Therefore, it is possible that the metal items were brought to the place as nomads' plunder, collected by those people along their routes and brought to the campsite as scrap metal for re-working. Sometimes casting was defective, so the failed castings were recycled again. All the stages of metallurgical production are represented in the finds from 'Uqdat Al-Bakrah: metalwork ready for recycling, new objects, casting residues, ingots, refining processing waste.

Compositional analyses carried out with X-Ray Fluorescence on a large selection of objects – more than 60 – revealed that only a small amount, about one quarter, was made of real tin bronze. The remainder was produced with almost pure copper. There are no objects made of arsenical copper: they generally have less than 1% As. Nickel was detected in most of the items, but in value less than 1%; one tenth only has a nickel content between 1% and 3%. Silver is absent or it has values below 1% (Giardino and Paternoster, forthcoming). These data suggest that Iron Age copper had a different origin from that used in some of the previous periods.

The hundreds of copper based finds from 'Uqdat Al-Bakrah prove not only that metal was easily available during the Iron Age in Oman but also that the country was one of the most relevant production centers in the Arabian Peninsula. The total amount of copper items from this site demonstrates a clear production surplus, even if the paucity of real bronze objects suggests that tin was still a rather rare metal in this period as well. 'Uqdat Al-Bakrah is far away from the large copper ore deposits that characterize large areas of the country. Nevertheless, the site was a very important metallurgical workshop, whose production was not only utilized by the local communities, but it was also widely exported.

There is no evidence at all of copper smelting at the site: the furnaces were devoted to copper melting and fuel production. Trade through the desert supplied the workshop of 'Uqdat Al-Bakrah with their metal requirement. Scrap metal was carried to the site for casting new objects. To avoid uneconomical waste of space during the transport, the dimensions of the old metalworks were reduced and carefully compacted by hammering (Figure 11.3). In this way, they were also immediately ready for re-casting.

The industrial area

'Uqdat Al-Bakrah was a large industrial area where a large spectrum of manufacturing activities was performed altogether. Here were found not only a large number of items related to metallurgy (i.e. stone and bronze tools, scrap, casting jet, casting residues, etc.), but also the remains of structures for metal production. These craft indicators allow us to examine the different stages of the metallurgical operational chain. The site is well preserved by its location, in the sands of the Rub Al-Khali, far away from possible damage, until now; the dry environment contributed to the good preservation of the metal items. It was possible to distinguish two main productive structures, one for fuel production and another for casting, refining and smithing activities.

Charcoal pits

The production of fuel is an important task for metallurgical activity; nevertheless, it is rarely recognized in prehistoric archaeological contexts. Charcoal is made by partial combustion of pieces of wood in limited air supply, and was frequently produced in pits. A special skill is required to produce good charcoal. Traditionally charcoal is produced in stacks covered with earth or else in pits (Craddock 1995: 191-196). The evidence from 'Uqdat Al-Bakrah suggests that here the pit method was adopted. Charcoal burning is attested by charcoal remains that were recovered at the bottom of large pit kilns, relatively deep: they were most probably devoted to charcoal (Figure 11.4). These pits were almost quadrangular or sub-ovoid in shape, more than 1 m long. Their size, bigger than an ordinary metallurgical furnace, is a clear indication of their use as charcoal pit kilns.

Prehistoric furnaces for metal smelting or casting were in fact about 40-60 cm in diameter, in order to avoid the dispersion of heat and the need for large amount of fuel. Sometimes charcoal pits were assembled together: groups of three quadrangular pits were recovered during the excavation. According to traditional techniques and archaeological experiments, wood was piled around a large pole (Kenny 2010; Kenny and Dolan n.d.).

Figure 11.3. 'Uqdat Al-Bakrah. Dagger folded up for recycling (# 207): A) frontal view; B) lateral view (photographs by C. Giardino).

Figure 11.4. 'Uqdat Al-Bakrah. Charcoal pit (photograph by C. Giardino).

Figure 11.5. 'Uqdat Al-Bakrah. Furnace with stones around the perimeter (photograph by C. Giardino).

Figure 11.6. 'Uqdat Al-Bakrah. Large casting residue that preserved as a cast the inside of a crucible (DA 27317 B (photograph by C. Giardino, modified by P. Koch).

Then the wood was covered with a thick layer of green bushes and leaves, and finally with soil that was, in this case, sand. The bushes prevented sand joining the charcoal. Finally, the pile was removed to create a chimney.

Burning charcoal was introduced from the chimney to ignite the wood. The production of good charcoal is crucial in any metallurgical activity: it should be considered that to heat a small furnace about 20 kg of charcoal were needed (Giardino 2010: 59- 61). The process of converting wood into charcoal takes from 14-16 hours to some days to produce ready-to-use lumps of charcoal, depending on the wood used and its quantity. The vegetation of the outer margin of the Rub Al-Khali (*Prosopis cineraria* and *Acacia ehrenbergiana*) alimented the production of charcoal, even if these plants were probably not abundant.

Furthermore, wood loses about 80% of its original weight during the carbonizing process. Therefore, it is possible that also other fuels, as camel dung, were used: ethnographic examples show that dung is still used in the tropics, together with charcoal to feed small forges (Craddock 1995: 196). The desert is an optimal environment for charcoal production: charcoal fires emit colorless, odorless and poisonous carbon monoxide (CO_2) gas, so they must be made in confined spaces and in a windy area, in order to disperse the gas.

Casting furnaces and forges

Casting furnaces were smaller structures than the charcoal pit kilns. They were used to obtain a small, well controlled environment where it was relatively easy to reach the high temperature required to liquefy the copper alloys (around 900-1000 °C), in order to cast the objects. The shape of the furnaces is hard to establish, because the surviving evidence is confined to the base, as frequently happens. They were circular or irregularly ovoid, with around a 60 cm inner diameter (Figure 11.5).

Sometimes, a circular line of stones underlined the perimeter, an indication that the furnaces had stone walls, probably with clay used as mortar. Probably small pit furnaces were also used as forges for iron smithing. In antiquity, pit furnaces were frequently used for iron smithing and for copper melting too. The presence of a smithing activity at the site is attested by the presence of semi-finished iron artefacts.

Casting activities

Copper alloys were melted inside crucibles, even if only a few fragments of clay crucibles were recovered during the excavation. Probably the high melting

Figure 11.7. 'Uqdat Al-Bakrah. Small casting residue (DA 26653.a) (photograph by C. Giardino, modified by P. Koch).

Early Iron Age, ca. 1300-300 BC

Figure 11.8. 'Uqdat Al-Bakrah. Semi-finished cast axes (photographs by C. Giardino, modified by P. Koch).

temperatures damaged the coarse pottery, made of refractory clay and small pebbles. A casting residue kept the shape of one of them; it gave an indication of a crucible shape and dimension (Figure 11.6). It was a shallow, hemispherical bowl of about 15-18 cm diameter, which could contain inside more than 1 kg of copper. Indicators of casting technology are many casting residues that were found scattered in the sand, as prills, sprues and cast scraps (Figure 11.7).

Many different copper-based objects were manufactured at 'Uqdat Al-Bakrah. Unfinished objects, as axes, were recovered at the site, probably left for recycling because of casting defects (Figure 11.8). Casting residues and sprues were collected in order to be re-cast. Sprues were cut from the cast during the finishing process; a sprue still had the preparatory line engraved on its surface, to ease detachment from the object. The annealing process is strictly connected with casting activity; it took place in the same forges/furnaces. It is a heat treatment that increases ductility of a metal and makes it more workable by heating it in a forge until the metal is dull red.

Stone tools

Casting did not only take place at 'Uqdat Al-Bakrah, but other processes too, all related to metal objects production. Many lithic tools emerged from the surface of the site: many stones had wear traces that testify they were used as hammers or anvils (Figures 11.9 and 11.10). Stone hammers were used in rough processes, like breaking the bronze objects before recycling or to flatten and to harden the bronze blades after casting. Whetstones were recovered at the site too (Figure 11.11).

They were used to sharpen the edge of weapons and instruments blades cast in the workshop. Whetstones are a relevant clue that a full production process took place at 'Uqdat Al-Bakrah because they provide evidence of finishing. Some argillite blocklets recovered at the site were probably used in the finishing process too: argillite, added to water or better to organic oil, produces a good polishing paste, thanks to their very fine, abrasive powder, to make metal artefacts shine.

The refining operations

According to the 'Uqdat Al-Bakrah evidence, some refining process took place far away from the smelting sites, in the casting workshops. The main Omani copper mineral was chalcopyrite ($CuFeS_2$), where copper is combined with iron and sulfur. Sulfur was eliminated by roasting the ore, but iron passed into the metallic copper during the smelting processes: therefore, the raw copper needed further purification before casting.

At 'Uqdat Al-Bakrah fragments of copper ingots were found, together with cast scraps containing a superficial iron patination, clue of a high iron content, as expected in unpurified copper (Figure 11.12). The analyzed ingots were made of refined copper, almost pure, while some of the scraps were so iron rich – sometimes they had around 40% of iron – that they attracted magnets. The copper had to be purified in order to remove the iron before being used in object production, just before alloying, because it could cause serious embrittlement. Silicatic sand was a useful ingredient for this refining and purifying operations: sprinkling sand upon the surface of an open crucible containing liquid copper produce the formation of slag that removes the iron and other impurities (Craddock 1995: 203-204). The Rub Al-Khali sands are composed of 80 to 90% quartz. Sometimes copper and iron oxides contained in the copper metal would react with the wall of the clay crucible to form small, reddish slag; one of these slags was found in the excavation.

Early Iron Age, ca. 1300-300 BC

Figure 11.9. 'Uqdat Al-Bakrah. Dimpled stone tools on the the sand.

Figure 11.10. 'Uqdat Al-Bakrah. Stone anvil on the surface of the site.

Figure 11.11. 'Uqdat Al-Bakrah. Pierced whetstone (#170); on the left corner, detail of wear traces.

Figure 11.12. 'Uqdat Al-Bakrah. Fragment of a plano-convex copper ingot (DA 29943) (photograph by C. Giardino, modified by P. Koch)

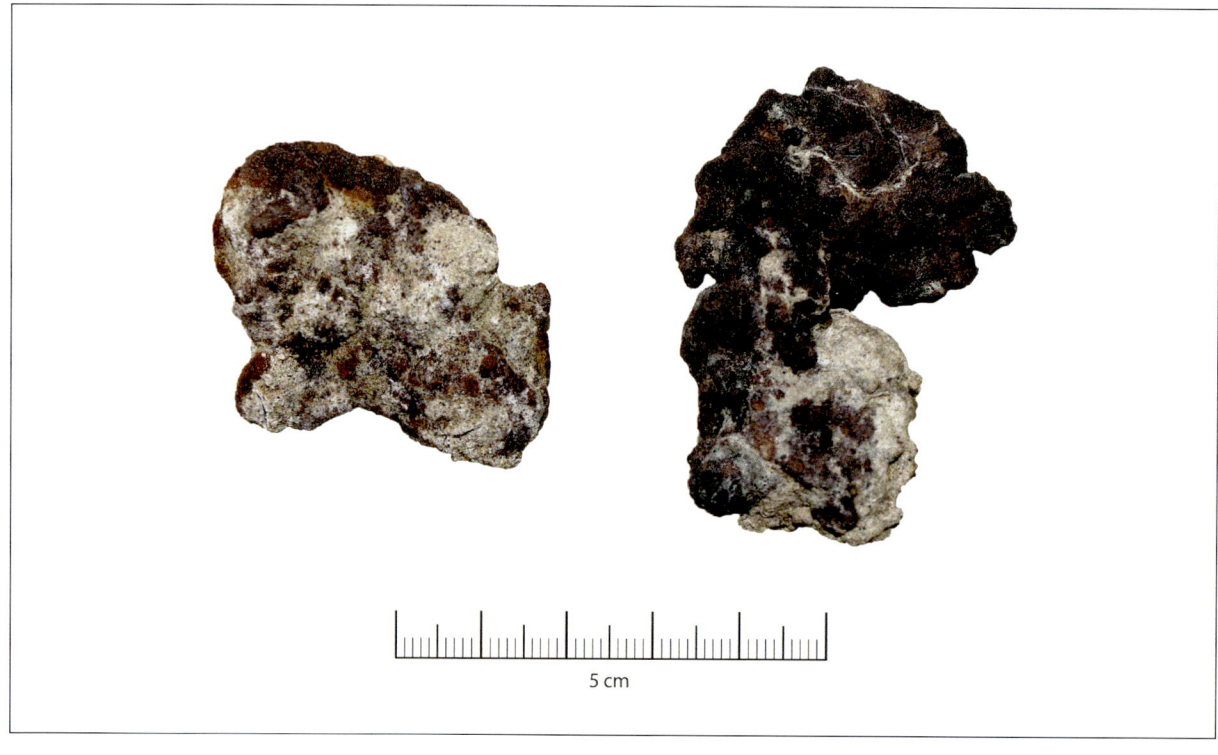

Figure 11.13. 'Uqdat Al-Bakrah. Iron blocklets (DA 26655.b) (photographs by C. Giardino, modified by P. Koch).

Iron metallurgy

'Uqdat Al-Bakrah is one of the earliest sites in Oman were traces of iron metallurgy were detected. Not only copper-based casting was verified at 'Uqdat Al-Bakrah, but iron working too. Some of the small furnaces found at the site could also work as forges. The use of the same structures for both iron forging and copper melting is not uncommon in ethnological and archaeometallurgical studies. Iron blocklets – a sort of small iron ingots – and semi-finished iron objects testified that iron objects (probably blades) were produced at the site (Figure 11.13). Iron could arrive to 'Uqdat Al-Bakrah as ingots or as bloom. Bloom is the first result of iron ore smelting; the main iron ores were magnetite (Fe_3O_4), hematite (Fe_2O_3), goethite ($FeO(OH)$), limonite ($FeO(OH).n(H_2O)$) and siderite ($FeCO_3$). In ancient Europe, there is evidence of bloom trade from the smelting places to the centers devoted to object production (Giardino 2011).

Typology of tools and vessels from 'Uqdat Al-Bakrah

At present 'Uqdat Al-Bakrah is the only Iron Age metalwork found in Oman. Therefore, the items that were found at the site are relevant not only to understand the typology of the objects of this period, but also to comprehend the metallurgical operations carried out in a prehistoric industrial site and to reconstruct its operative chain. The finds that were probably used in many, different stages of metallurgical production will be examined in the following pages, from the axes for wood cutting to make charcoal, to the hammers to shape the casts. It should be noted that a large amount of the metal objects was made of unalloyed copper. Tin bronze was observed only in 1/3 of a large group (more than 60 items) of objects analyzed with XRF; it was mostly reserved for the production of axes and weapons.

Axes

Axes are an ambiguous class of archaeological materials, because these artefacts could have many different uses. Shaft hole axes similar to 'Uqdat Al-Bakrah items were recovered inside Iron Age tombs, as at Rumaylah (UAE), at Quasys (Tomb Qu24) and at Qarn Bint Saud (Tomb QBS3), and in the Selme hoard.

A stone seal from Rumaylah represent a man brandishing an axe with a cutting edge turned against him: it was interpreted as a cult related scene (Lombard 1969: 72; Yule and Weisgerber 2001: 18, figs. 9). Axes were at the same time weapons, but also tools: in a complex metallurgical place like 'Uqdat Al-Bakrah the axe was an integral part of the workshop tool kit, essential for many purposes, as for cutting and splitting the wood devoted to charcoal production. Pit kilns found during the excavation testify that making charcoal was a relevant activity at the site (Figure 11.14).

A very peculiar group of axes is A6, because they differ from all the other ones not only for their characteristic shape, but also because they were not made by casting. They were produced starting from broadly lozenge-rectangular thick blades whose smaller end was rolled up (probably around a wooden or metal bar) by hammering in order to produce an eye for handling; the cutting edge was made by mechanical deformation too. These axes are very uncommon in archaeological contexts of Southeastern Arabia. One of them was recovered at the metallurgical Early Iron Age site of Saruq Al-Hadid, where it was supposed that it could have been used in manufacturing processes (Qandil 2005: 133, fig. 11.3).

A1 – Slim shaft-hole axes with curveted and parallel sides.
[DA 27304A (ceremonial axe with representations of bearded head and lion)].

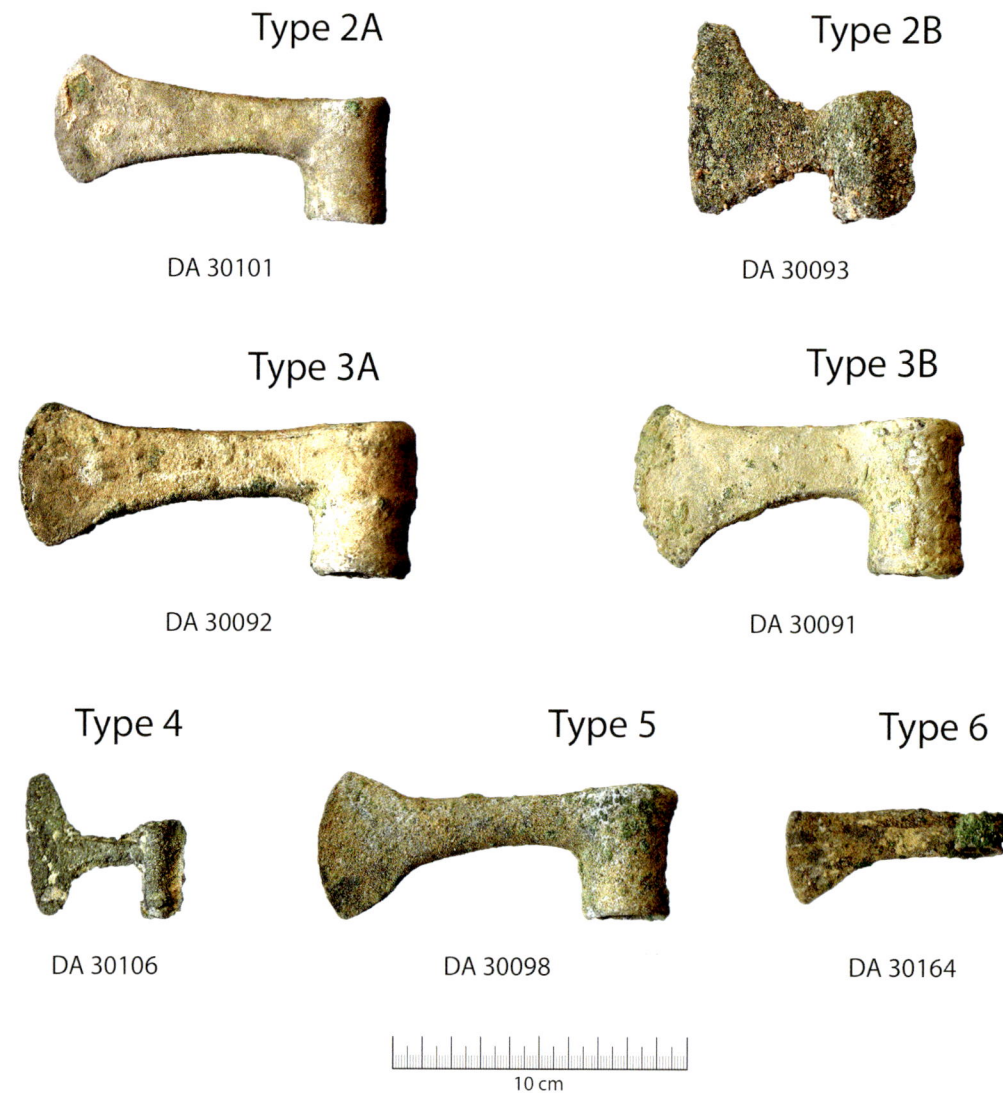

Figure 11.14. 'Uqdat Al-Bakrah. Typology of axes (photographs by C. Giardino, modified by P. Koch).

A2A – Slim shaft-hole axes with diverging sides, the upper one curveted, the lower one slightly curveted.
[DA 30102; DA 27302; DA 30100; DA 30087; DA 30096 (blade fragment, with sprue); DA 30090; DA 30088; DA 30099; DA 30095; DA 30101].

A2B – Thick shaft-hole axes with diverging sides, the upper one curveted, the lower one slightly curveted.
[DA 30103 (cast scrap, with sprue and casting seam); DA 30093 (small); DA 30094 (small); DA 30163 (small)].

A3A – Slim shaft-hole axes with diverging sides, the upper one straight and the lower one curveted.
[DA 30092; DA 30097; DA 30089; DA 27304].

A3B – Thick shaft-hole axes with diverging sides, the upper one straight and the lower one curveted.
[DA 30091; DA 29670 (with triangular shaft)].

A4 – Shaft-hole axes with diverging sides, both heavily curveted.
[DA 30106; DA 30105 (cast scrap)].

A5 – Shaft-hole axes with parallel, horizontal sides.
[DA 30098; DA 30104].

A6 – Shaft-hole axes with diverging sides, the upper one straight and the lower one heavily curveted; not symmetric. These items differ from all other axes for their manufacturing process – the eye for handling was not made by casting, but by rolling up the butt by hammering.
[DA 29660; DA 27315; DA 30164; DA 30240; DA 30252].

Hammers

Metal hammers were used in the 'Uqdat Al-Bakrah workshop together with stone ones. The latter were mostly employed for heavy work, where blow precision did not matter. Copper-based hammers vary in size, shape and weight, depending on their purpose (Figure 11.15). Weight is in fact a relevant characteristic in a tool used to deliver blows to an object, mostly to shape it. Some of them (for example 1A, 3A, 4) were ball-peen hammers, a kind of hammer used in metalworking for hardening metal surfaces by impact, for rounding off edges of artefacts and for raising vessels. According to their peculiar shape, the hammers from 'Uqdat Al-Bakrah were also useful for raising metal vessels: the different hammer sizes could be used on pieces of different size and different thickness of metal. H1B and H3B are planish hammers: they have two distinct profiles, one end is convex, which is used for the first planishing, and a flat end used to finish planishing (cf. Maryon 1971: 89-92). H6 is also of special interest. The typical use of hammers of this shape was to harden the vessel walls: traditional artisans still use this very specialized tool and its use in antiquity is testified by its representation in a Greek painted vase from Borgo di Gela in Sicily dated to the 6th century BC (Giardino 2008: 28-29; Orsi 1906: 54-56).

H1A – Triangular shaped hammer, with two rounded faces.
[DA 30174 (weight g 186); DA 30282 (weight g 151); DA 30283 (weight g 275); DA 30280 (weight g 163); DA 30276 (weight g 121); DA 30179 (weight g 174)].

H1B – Triangular shaped hammer, with one rounded face and the other flat; round hafting tube.
[DA 30177 (weight g 90); DA 30283 (weight g 107); DA 30223 (weight g 305)].

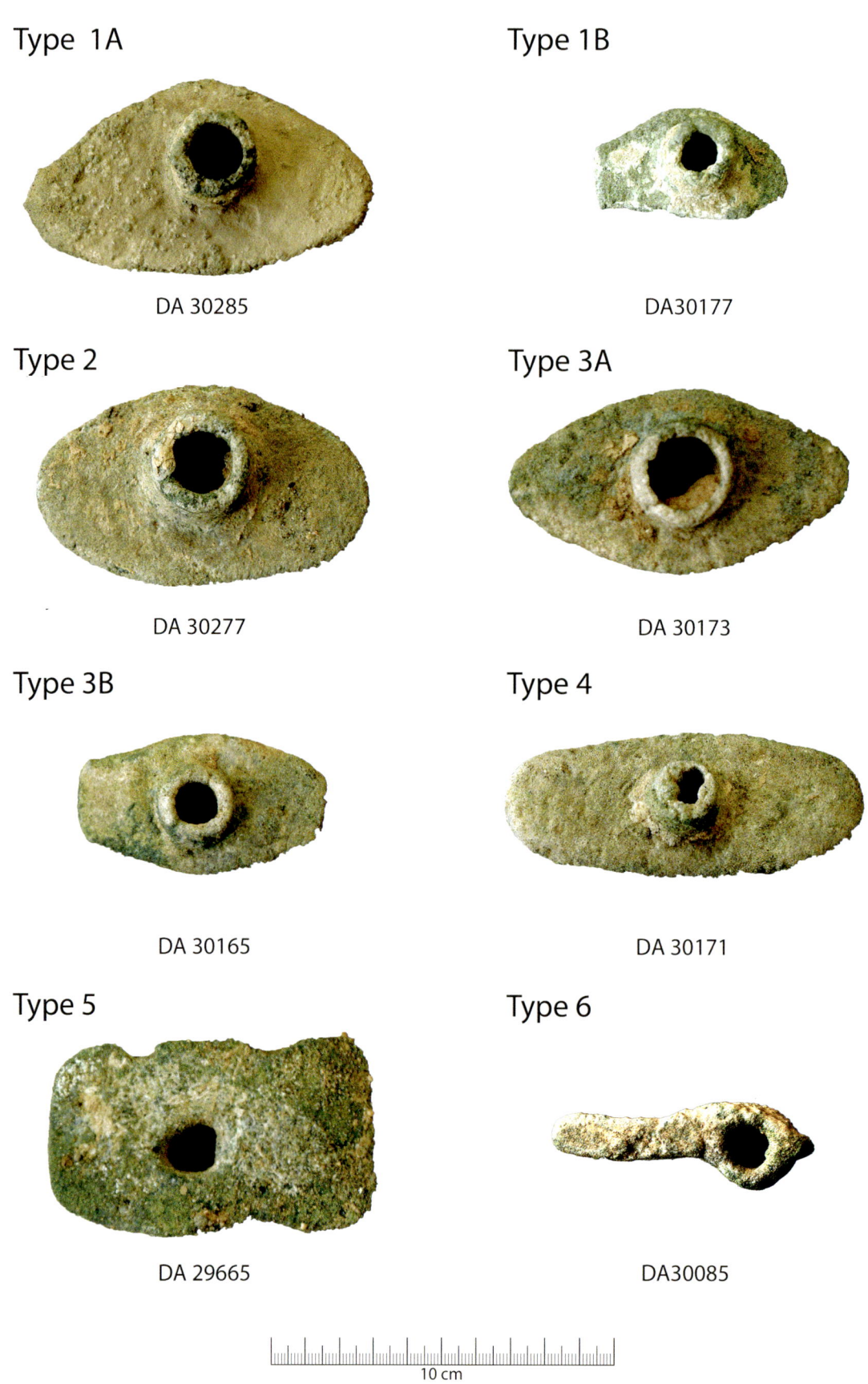

Figure 11.15. 'Uqdat Al-Bakrah. Typology of hammers (photographs by C. Giardino, modified by P. Koch).

H2 – Oval shaped hammer; round hafting tube.
[DA 29659 (weight g 338); DA 30278 (weight g 252); DA 30281 (weight g 207); DA 30277 (weight g 253); DA 30176 (weight g 86); DA 30279 (weight g 249)].

H3A – Rhomboid shaped hammer, with rounded faces; round hafting tube.
[DA 30173 (weight g 215); DA 30175 (weight g 213)].

H3B – Rhomboid shaped hammer, with flat faces; round hafting tube.
[DA 30165 (weight g 130)].

H4 – Rectangular shaped hammer (raising hammer); round hafting tube.
[DA 30178 (weight g 64); DA 30171 (weight g 220); DA 30172 (weight g 64)].

H5 – Flat shaped hammer (collet hammer); round hafting tube.
[DA 29665 (weight g 273); DA 30166A (weight g 266)].

H6 – Shaft-hole hammer with square, long peen (hardening hammer).
[DA 30085 (weight g 85)].

Stakes

Stakes are special tools for raising and shaping metal vessels, characteristic of bronze craft, over which a sheet metal is hammered to create a hollow form. They can have many different shapes. One of the most typical and useful for raising is the T-stake: one end is square cut for working on hollow forms, which require a flat bottom, the other end is rounded to use on round bottoms. Raising is the shaping of metal from a flat sheet into a vessel by using special hammers, stakes, and proper placement of hammer blows (Maryon 1971: 87-89). The 'Uqdat Al-Bakrah stakes are elliptical in section; generally, one end is flattened on top and the extremity undercut to make an angle (Figure 11.16).

S1 – Long stakes.
[DA 30082; DA 30080; DA 30081; DA 27317; DA 30079; DA 30269; DA 30266; DA 30263; DA 30253; DA 30254; DA 30251; DA 30079].

S2 – T stakes.
[DA 30083].

Shaft-hole hoes

Shaft-hole hoes are versatile agricultural tools used to dig or move soil. The hoes from 'Uqdat Al-Bakrah have parallels in several places of the Arabian Peninsula, even if so many items were never recovered in a single site. Similar hoes were found in Oman at Ibri and Sachrut Al-Hadri on the Masirah island; at Rumaylah in the UAE and at Al-Midra Ash-Shamali in Saudi Arabia (Weisgerber and Yule 2003: 46, figs. 28: 1, 3, 4; 29: 1, 2) (Figure 11.17). In this specific metallurgical context hoes were used to dig the pits for making furnaces and kilns, but also in any operation which needed sand shifting.

Ho1 – Rectangular shaped hoes with round hafting tube.
[DA 30167; # 255; DA 30163; 364 P; 2013/379 P].

Figure 11.16. 'Uqdat Al-Bakrah. Typology of stakes (photographs by C. Giardino, modified by P. Koch).

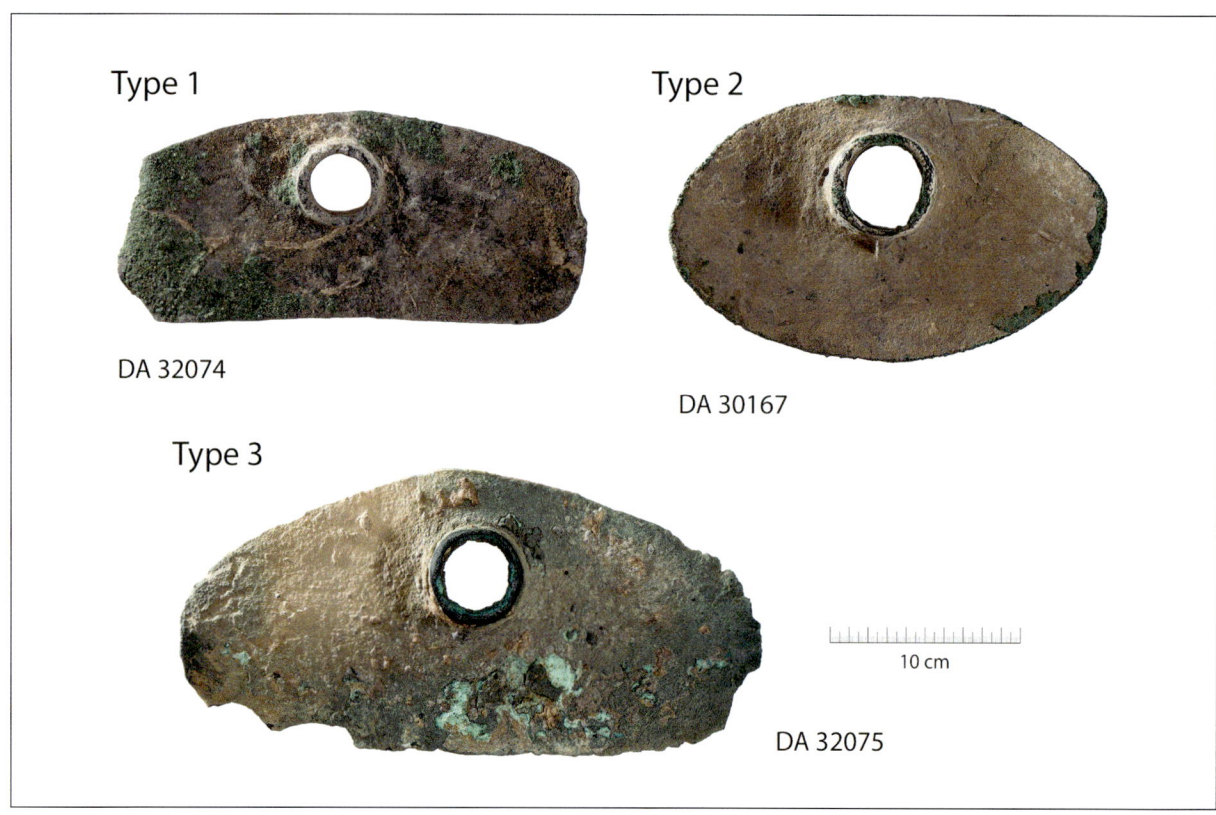

Figure 11.17. 'Uqdat Al-Bakrah. Typology of hoes (photographs by C. Giardino, modified by P. Koch).

Ho2 – Ovoid shaped hoes with round hafting tube.
[DA 30168; 2013/82 P; 2013/03 P; 2013/174 P].

Ho3 – Triangular shaped hoes with round hafting tube.
[2013/412 P].

Razors

Razor blades are not common in Oman; this is the largest assemblage of these objects found all together in the country. During the Early Bronze Age razors had an oval or quadrangular shape with a short tang. In the Iron Age, they assumed a trapezoid form; the items from 'Uqdat Al-Bakrah have a parallel in the contemporary grave from Nizwa (Bakhit Al-Shanfari and Weisgerber 1989: 20, fig. 2.4) (Figure 11.18).

In a metallurgical site, strong and sharp blades had a role not only for cutting, but also for leveling, spreading, or shaping substances such as plaster, in order to build or repair furnaces and kilns, or to mix together powders and other ingredients such as modern trowels.

R1 – Slim razor with straight sides and flat or slightly curveted heel.
[DA 30216; DA 30203; DA 30214; DA 30215; DA 30207].

R2 – Large razor with straight sides and flat or slightly curveted heel.
[DA 30213; DA 30312; DA 29958A; DA 30201; DA 30205; DA 30211].

R3 – Slim razor with straight sides and curveted heel.
[DA 30208; DA 30200; DA 30209].

R4 – Slim razor with curveted sides and flat or straight heel.
[DA 30204].

R5 – Slim razor with heavy curveted sides and flat or straight heel.
[DA 30202; DA 30210; DA 30187].

R6 – Rectangular shaped razor, with parallel sides.
[DA 30199].

R7 – Thicker, trapezoidal razor.
[DA 30198].

Spatulas

A spatula is a small, light tool with a flat, flexible blade used to mix, spread and lift materials such as foods, plaster and colors, but also to scrape or transfer powders and dense liquids. In a metallurgical workshop, spatulas could be useful to stir or to skim hot, molten metal as well (Figure 11.19).

S1 – Tanged spatula with broad blade.
[DA 30067].

S2 – Spatula with small blade and long tang.
[DA 30192; DA 30193].

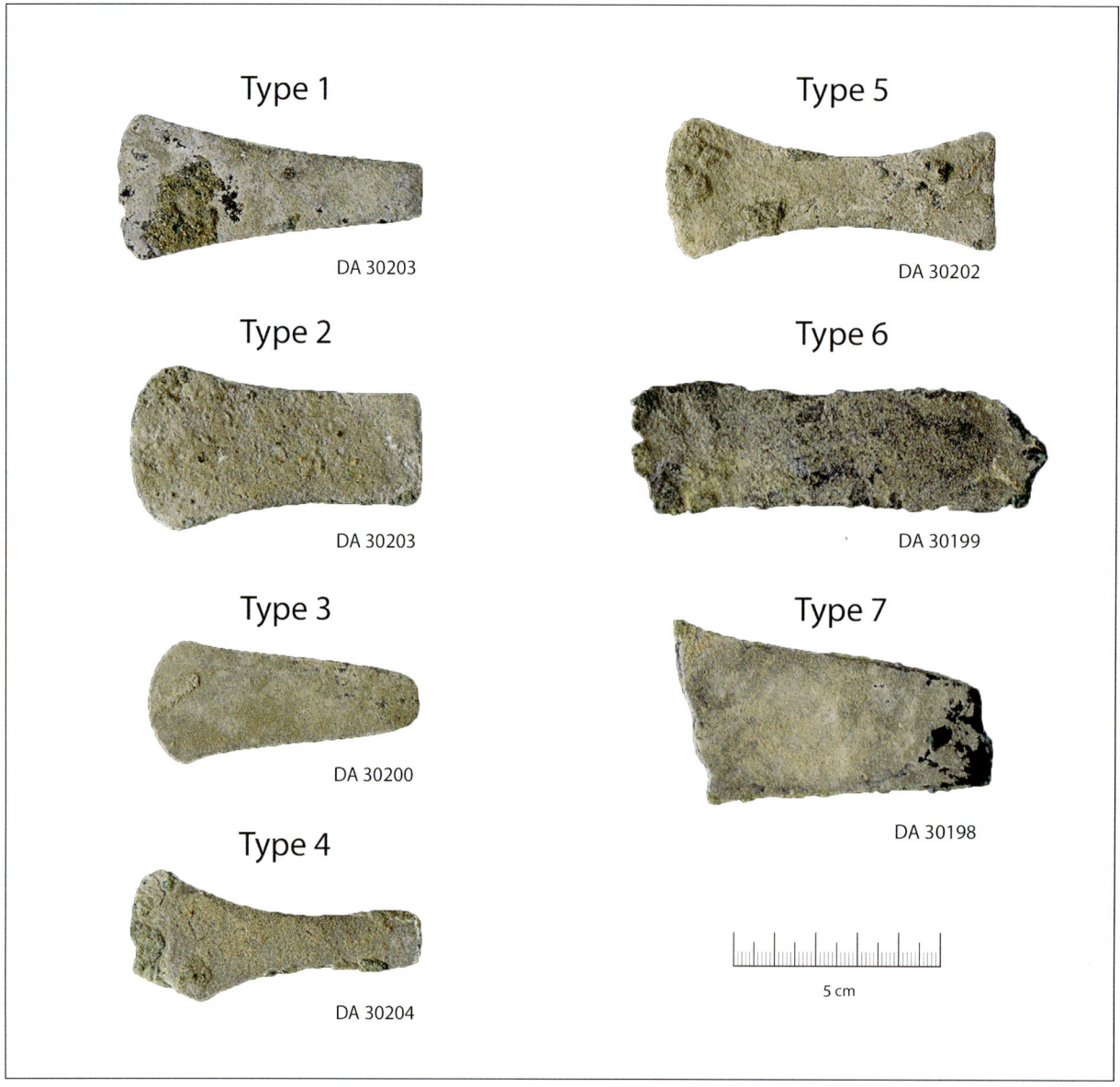

Figure 11.18. 'Uqdat Al-Bakrah. Typology of razors (photographs by C. Giardino, modified by P. Koch).

Chisels

Chisels are not very common tools at 'Uqdat Al-Bakrah. They were used for engraving, in order to decorate the surface of objects with a sharp point. Generally, chisels were manufactured with copper alloys, but they were also made of iron, probably to engrave hard, bronze artefacts in an easier and more efficient way (Figure 11.20).

C1 – Copper based chisel.
[DA 30196; DA 27322].

C2 – Iron chisel.
[DA 30194].

Vessels

In the Iron Age, Southeastern Arabia metal vessels were intended for eating, drinking, preparation of meals and toiletry use (Yule and Weisgerber 2001: 39). Some of the small, shallow bowls were used as scale plates: during the excavation, a couple of these plates were found joined together by corrosion (Figure 11.21). Many copper-based bowls were recovered during the excavation at 'Uqdat Al-Bakrah; bowls are mainly divided into shallow (A) and steep-walled bowls (B) (Figure 11.22). Vessels were generally made of unalloyed copper, to make the raising process easier. Liquids as water were not only useful in metal production to control fires, but they were also necessary for the quenching process, the rapid cooling of a metal workpiece to obtain certain material properties. In copper alloys, a water quench after annealing allows obtaining a more malleable metal than just an air quench.

A – Shallow bowls

A1 – Shallow bowls, hemispherical or lightly truncate-conical shaped, spouted (A1):
[DA 29867; DA 29874; DA 29869A; DA 29871; DA 27326; DA 27328; DA 29871; DA 29994; DA 29985; DA 29996; DA 29988; DA 29984].

A2 – Shallow bowls, hemispherical or lightly truncate-conical shaped not spouted (A2):
[DA 29869A; DA 29870; DA 29875; DA 29873; DA 29872; DA 29868A; DA 29868B; DA 29885; DA 29886; DA 29990; DA 29876; DA 29995; DA 29993].

B – Steep-walled bowls

B1 – Steep-walled bowls, truncate-conical shaped, not spouted (B2):
[DA 29878; DA 29877; DA 29880; DA 30000].

B2 – Steep-walled bowls, hemispherical shaped, spouted (B1a):
[DA 29879; DA 29881].

B3 – Steep-walled bowls, globular shaped, spouted (B1b):
[DA 29879; DA 29881].

B4 – Steep-walled bowls, truncate-conical shaped, spouted (B1c):
[DA 29887; DA 29987; # 193].

Figure 11.19. 'Uqdat Al-Bakrah. Typology of spatulas (photographs by C. Giardino, modified by P. Koch).

Figure 11.20. 'Uqdat Al-Bakrah. Typology of chisels (photographs by C. Giardino, modified by P. Koch).

Figure 11.21. 'Uqdat Al-Bakrah. Pair of shallow bowls used as scale plates (DA 29869 A, B) (photographs by C. Giardino, modified by P. Koch).

Early Iron Age, ca. 1300-300 BC

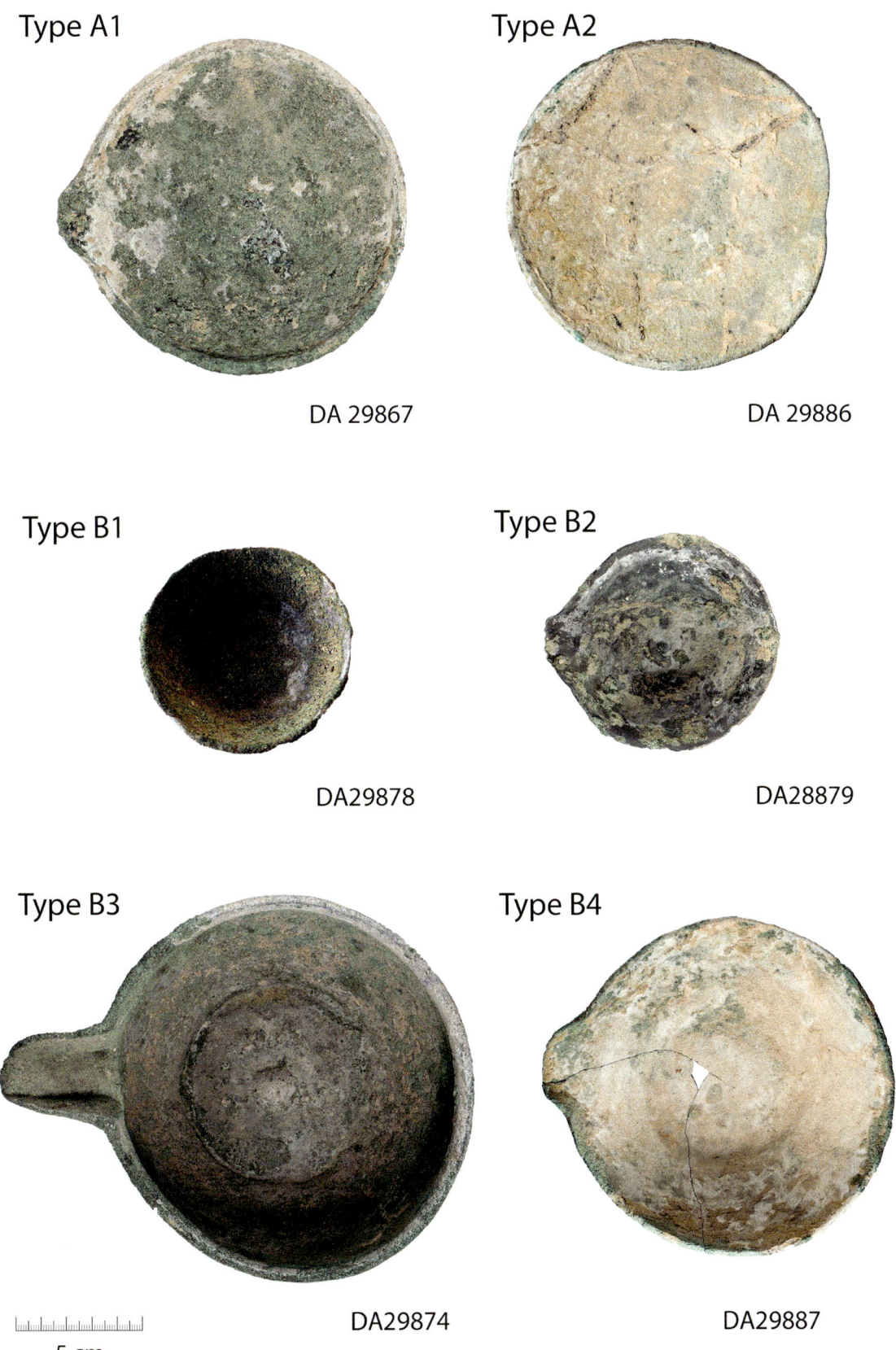

Figure 11.22. 'Uqdat Al-Bakrah. Typology of vessels (photographs by C. Giardino, modified by P. Koch).

Precious metals

At 'Uqdat Al-Bakrah a few jewels were also recovered: they are an indication of the wealth of the people who worked at this place. Gold and silver objects were found on site. Silver was used to produce an elegant finger ring (DA 30197): it has an empty socket that contained a lost stone (Figure 11.23). Special interest has a carnelian bead mounted in gold, with a band decorated with granulation (DA 26169) (Figure 11.24). This is a sophisticated jewelry manufacturing technique whereby a precious surface is covered in microscopic spherules of gold. It appeared in Mesopotamia as early as the middle of the 3rd millennium BC (Giardino 2010: 100-104).

Figure 11.23. 'Uqdat Al-Bakrah. Silver finger ring (DA 30197) (photograph by C. Giardino, modified by P. Koch).

Iron Age II (Lizq period): the Selme hoard

The large hoard of Ibri/Selme is chronologically slightly later than the previous evidence from Nizwa, Daba and 'Uqdat Al-Bakrah (Figure 11.25). The hoard is a good example of Omani Iron Age II metal production, that can be dated to the first part of the 1st millennium BC, the so called Lizq period. It contained 550 metal objects, including weapons and metal vessels, frequently crushed: it is the largest hoard in the ancient Near East (Weisgerber and Yule 1989; Yule and Weisgerber 2001). This assemblage had a huge amount of metals: some of the bangles recovered in the hoard weighed more than 600 g each. Many bangles were contained in the hoard; bangles were a rather common ornament in the Iron Age that belonged to men, women and children too.

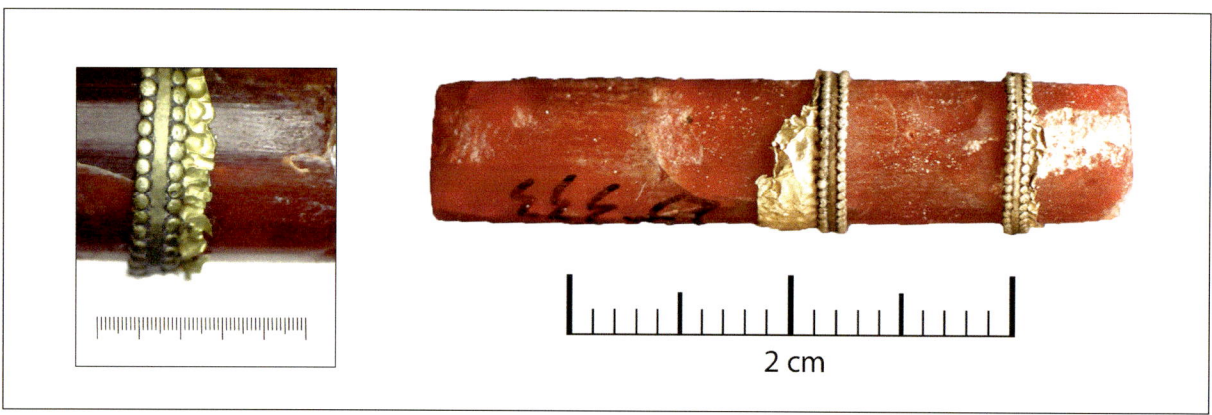

Figure 11.24. 'Uqdat Al-Bakrah. Carnelian bead decorated with gold granulation (DA 26169.m). Top-left corner: micrograph of the granulation (photographs by C. Giardino, modified by P. Koch).

Early Iron Age, ca. 1300-300 BC

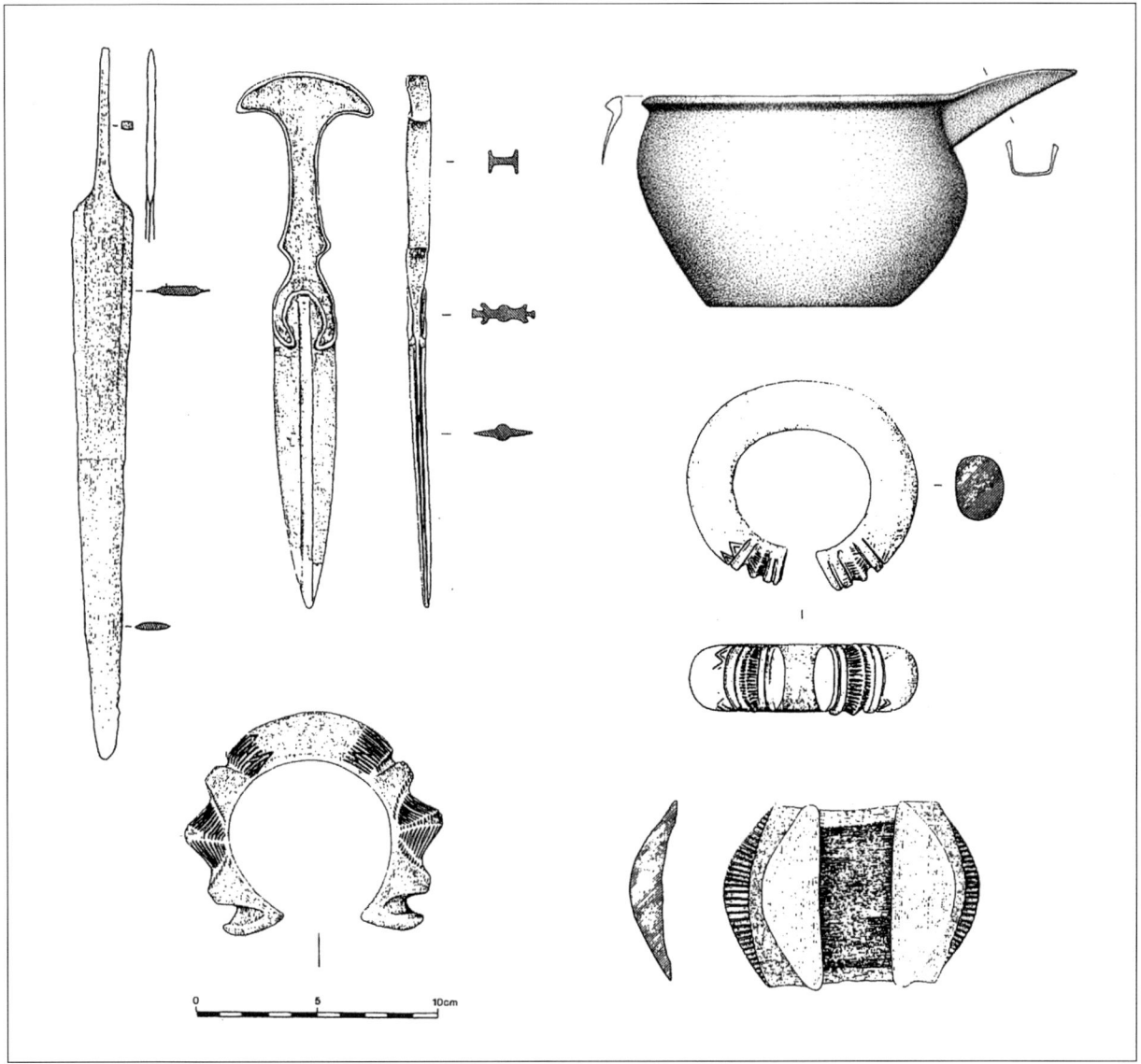

Figure 11.25. Metal objects from the hoard of Ibri/Selme (after Weisgerber and Yule 1989; Yule and Weisgerber 2001).

The hoard was found at Selme near Ibri, in a desert plain, and it was probably the cache of ancient grave looters that sacked Lizq period tombs: it was mainly found in an Umm an-Nar type grave that was reused for storage. Grave robbing had to be a systematic activity during the Iron Age, as an alternative method for the acquisition of metal instead of mining and smelting. The evidence from Daba shows the huge richness of a single burial. Probably the activity of the 'Uqdat Al-Bakrah workshops was alimented not only by trade, but also by grave plundering. Other objects found were, daggers, arrowheads, vessels, massive bangles and anklets.

Copper vessels presented a large variety of shapes, mainly globular open bowls, open bowls with a flat bottom and open rim, and carinated bowls. Different techniques were used for the production of the metal objects from Selme: casting and smithing to shape daggers and bangles, raising (and perhaps castings for only one item) for vessels. Metallurgical studies carried out on a bangle show that it was not just cast, but it was submitted to cycles of cold and hot workings before reaching its final shape (Prange and Hauptmann 2001: 80, fig. 24).

Some bangles and some vessels were decorated by an incision. No iron items were recovered in the hoard, probably because it was not possible to recycle iron. 86 items from the Selme hoard were analyzed by Inductively Coupled Plasma Optical Emission Spectrometry (ICP-OES) and Atomic Absorption Spectrometry (AAS) (Prange and Hauptmann 2001: tab. 9). According to these analyses, almost all objects were manufactured in tin bronze, containing about 7-13% Sn; average is 9.9% Sn. Only one find was not made in bronze: DA 5631 (<0.006% Sn), a tanged dagger. Another had a very low tin content, DA 3737 (1.55% Sn), an ovoid bangle open to one side. The only examined ax had a tin content of 7.8%, while weapons had a tin content up to 11% and bangles up to 13%. Looking at the arsenic and nickel, the bronze artefacts from Selme show a positive correlation with the Omani ore for nickel content (Prange and Hauptmann 2001: fig. 21). The metals also had a low iron content (between 0.01 and 0.1% Fe), an indication that copper underwent a purification process.

Chapter 12

Chemical-physical analyses by Energy Dispersive X-Ray Fluorescence (EDXRF)

Claudio Giardino and Giovanni Paternoster

In the following pages and tables, we present the results of XRF analyses carried out by X-Ray Fluorescence on the prehistoric metal objects from Oman. We analyzed, by Energy Dispersive X-Ray Fluorescence (EDXRF) with a portable apparatus, 265 bronze finds and 11 crucibles.

X-Ray Fluorescence analysis (XRF) was chosen primarily for its capability to perform measurements in a non-destructive way, i.e. without the need of taking out samples from the finds. The apparatus consists of an X-Ray tube (source), an X-Ray solid state detector and a Multichannel interfaced with a PC through a dedicated software (Figure 12.1). A helium gas flow system through the collimator of the X-Ray detector was used to be able to investigate, with greater sensitivity, a wider range of elements. The helium gas flow in front of the detector, replacing the air with an atmosphere of helium, improves the detection of low energy X-Rays, the K-lines emitted by the lighter elements and the L and M-lines from the heavier ones. The apparatus can reveal all the elements from potassium (K) and, when using helium (He) gas flow, it can also reveal those emanating from aluminum (Al). The characteristics of the apparatus and the operating conditions were, respectively:

- X-Ray tube with Pd (palladium) anode: supply voltage 38 kV max, current 300 mA max;
- Si-PIN detector with a resolution of 160 eV @ 5.9 keV;
- Geometry of irradiation: $\varphi X = 65°$ (the angle between the sample surface and the axis of the X-Ray beam), $\varphi D = 65°$ (the angle between the sample surface and the axis of the detector);
- Measuring live time: 200 s for bronze alloy, 300 s for crucibles;

The electronic signals from the detector are processed by a Multichannel Analyzer (MCA) and stored in a PC. The X-Ray Fluorescence spectra were analyzed with a commercial software package (Win-Axil™). The concentrations of the elements were calculated by the Fundamental Parameter method, using several standards to determine the system's calibration. The measured elements were: Al, Si, S, Cl, K, Ca, Ti, Cr, Mn, Fe, Co, Ni, Cu, Zn, Se, As, Rb, Sr, Y, Zr, Ag, Sn, Sb, Ba, Pb. Three certificate standards (IAEA SL-1, IAEA Soil-7 and IAEA SDM7) were analyzed with the same procedures to calibrate the system for the quantitative analysis of the crucibles. Moreover, five certified alloy standards (BCR-861) and six bronze alloy, locally made, were used for the calibration of the bronze alloy quantitative analysis.

The software package consists of two main programs. The first software performs the spectrum fit to give the area, with its standard deviation, of the peaks of the characteristic X-Rays of the elements which constitute the sample. The second one calculates the chemical composition in elements and/or chemical compounds. The results of the crucibles analysis are to be considered semi-quantitative, because the deposition of metal residues alters the concentration values of the matrix; on the other hand, our goal was just to reveal those metal residues. Therefore, we report the results using the cross symbol. In the tables, we report the MDL (Minimum Detection Limit), i.e. the minimum amount of elements that can be detected; moreover, we report the MRE (Minimum Relative Error), i.e. the minimum relative error we achieved, the maximum achieved at the MDL is 33%.

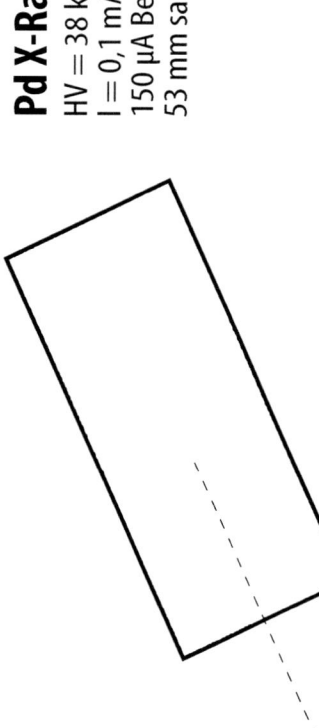

Figure 12.1. Schematic plan of the ED-XRF portable apparatus.

Analyses of the copper based objects

For the copper based finds we report the normalized concentration of the alloy, leaving in the text eventual information about abnormal presence of other elements connected with patina residues. In fact, it was not possible to thoroughly clean the patina from all the finds. It was not possible especially for the items characterized by a small section or very heavy corrosion, as it occurred generally in hooks, awls, punches or pins.

The sites of Ras Al-Hadd HD-6 and Ras Al-Jinz RJ-2 are located on sandy beaches close to the sea. Therefore, we detected amounts of calcium and chlorine in almost all the items. Moreover, silicon was seen in the most corroded or encrusted objects. Other elements such as strontium, titanium, manganese, potassium and chromium were detected only in trace amounts; furthermore, only an occasional presence of these elements was observed. Ten elements were taken into account to characterize the alloy contents of: copper (Cu), silver (Ag), zinc (Zn), iron (Fe), arsenic (As), cobalt (Co), nickel (Ni), lead (Pb), antimony (Sb) and tin (Sn) (Tables 12.1-12.6).

Special interest was devoted to the beginnings of Omani metallurgy, testified by the small finds recovered at the sites of Wadi Shab GAS-1 and Ras Al-Hamra RH-10: 79 objects belonging to this early period were examined. We analyzed 39 objects dated to the Hafit Period: they came from Ras Al-Hadd HD-6 (37 items), Ras Al-Jinz RJ-6 (1 piece) and Ras Al-Hadd HD-10 (6 pieces). A lot of the examined finds are dated to the Umm an-Nar period: 70 items; they were found at Ras Al-Jinz RJ-2. Analyses were also carried out on six metal items dated to the Wadi Suq period coming from Ras Al-Jinz RJ-1.

The histogram in Figure 12.2 shows the frequency/presence of elements in the main sites: frequency/presence represents the percentage of artefacts from the different sites in which considered elements exceed MDL (Minimum Detection Limit). We notice that silver, antimony and tin are lacking from the few finds from RJ-1; while the items from HD-6 have smaller amounts of cobalt, antimony, tin, and especially lead compared with those from RJ-1. More silver was generally detected in the earliest finds, from Wadi Shab GAS-1 and Ras Al-Hamra RH-10, than in the objects from later periods. Lead frequently occurs in RJ-2, while it is rare in GAS-1, RH-10, HD-6 and RJ-1 objects. The comparison between the earliest finds (GAS-1 and RH-10) and the Hafit items (HD-6, HD-10 and RJ-6) shows that the earliest objects have a similar behavior for arsenic, nickel, lead, antimony and zinc, but they frequently contain much more silver and cobalt: a possible indication that the copper could come from different sources.

The analyzed finds were divided into four categories: tools, ornaments, weapons and semi-finished objects. However, it should be stressed that, in examining the relationship between the type of objects and their elemental composition, the statistics could be flawed by an excess of the category "tools" compared to other items.

Copper (Cu)

Copper content is usually higher than 95%. A large number of Bronze Age metal objects from Oman are therefore made of almost pure copper (Figure 12.3). The oldest pieces, from GAS-1 and RH-10, normally have a higher copper content, between 98-99%; this is a probable indication that they were produced with native copper. About 20 items have a percentage lower than 95%, but above 90% Cu. Only two specimens from Ras Al-Jinz RJ-2 (a blocklet DA 12735 and a razor DA 12733) have copper contents lower than 80%. The first one has a high lead, arsenic and nickel content (18.2% As, 17.2% Pb, 2.6 Ni): it is therefore a leaded arsenic-copper alloy; the second one is a tin bronze with a relevant amount of tin (16.4% Sn).

Silver (Ag)

A relevant silver content characterizes the earliest pieces, from GAS-1 and RH-10, that frequently have 1-3% Ag, up to about 10% (Figure 12.4). This is a possible indication that the copper came from different sources in the earliest times of Omani metallurgy. Silver was detected in about half of the more recent finds; it seems to be independent from typology or chronology. In these objects, silver content is always lower than 1% (0.3-0.4% on average, about twice the MDL), except in a find from Ras Al-Hadd HD-6, an awl (or pin) DA 2247 (2.44% Ag); in this item silver is associated with a percentage of antimony (4.06% Sb). Because most probably the silver content is linked to the ore provenance, it is possible that the copper for the item from Ras Al-Hadd HD-6 had a different origin too.

Zinc (Zn)

Values around 0.15-0.30% in zinc characterize almost all the items, a percentage slightly higher than the MDL (0.06%); this content is not related to chronology (Figure 12.5). Using XRF, the determination of small quantities (less than 0.3%) is not very precise in alloys of almost pure copper because of the high copper signal, which overlaps the zinc signal, which is close to it. However, a similar concentration was also observed in previous analytical studies on finds from Southeastern Arabia that used different, more sensitive techniques for this element (Weeks 2003: 85).

Iron (Fe)

Iron was detected in most of the finds with values below ca. 0.5% (Figure 12.6). Higher contents are generally associated with a higher calcium content: in these cases, the greater amount of iron is due to the patina, because it was not possible to clean better the surface of that items. A good example of this is the Hafit dagger found in a grave from Ras Al-Hadd HD-10; the rivet of the dagger was small and heavily corroded: it shows an iron content of 5.60%, associated with a large amount of calcium. Iron content was below the MDL in three of the samples; two of these objects were made of almost pure copper; they were a blocklet (DA 11590) and a hook (DA 16981) from Ras Al-Hadd HD-6 and another blocklet from Ras Al-Jinz RJ-2 12735 made of a leaded arsenic-copper alloy. A pin from Ras Al-Jinz RJ-6 (DA 8621) also had a very low iron content (0.02%). A casting residue from the Late Bronze Age site of Ras Al-Jinz RJ-1 (DA 14405) had a relatively high iron content: 1.42%. Probably this residue is connected with a refining process that took place at the site in order to remove the iron from the raw copper before casting. The hook from Ras Al-Jinz RJ-2 (DA 10959) is completely made of iron. The analyses were carried out not only on the surface, but also on the section (the piece is broken). It is an exceptional find, unique in Oman because of its very early chronology.

Cobalt (Co)

Cobalt was present in trace amounts of about 0.02-0.03% in most of the earliest finds, while it reached values of 0.05% in about half of the other items (Figure 12.7). Only three items had a content that was much higher (0.16-0.25% Co); they all came from Ras Al-Jinz RJ-1 (DA 14405, 14406, 14313, respectively a casting residue, a hook and a needle). The presence of higher cobalt contents in copper-based objects dating to the Wadi Suq period was observed at Masirah (Site 38), Shimal (settlement Area SX and tomb 2) and Sharm (Hauptmann et al. 1988; Weeks 2000, 2003; Craddock 1985).

Arsenic (As)

Arsenic occurred in all specimens with contents ranging up to 4% (Figure 12.8). Most of the finds from GAS-1 and RH-10 had a low arsenic content, around 0.2-0.3%, too low to be considered an intentional addition. Only a drill point from GAS-1 had a significant amount of arsenic: 3.3%; two pointed tools and an awl from RH-10 reached 5.8%, 3.9% and 3.5% As respectively. There is no significant difference between the various classes of materials from RJ-2 for As content, while tools seem to mainly have less arsenic at HD-6 (Figures 12.9-12.11). The only exception was blocklet DA 12735 from Ras Al-Jinz RJ-2 that had 18.2% As. The diagram in Figure 12.8 shows two main item peaks: one has an As average content around 2-3%, the other one between 0.3-0.5%. The earliest finds (GAS-1 and RH-10) have only the second peak (0.3-0.5%), but this value occurs in the rest of the analyzed items, without chronological variations. These data agree with the previous analyses from Southeastern Arabia (Weeks 2003: 85-88). However, some objects could occasionally reach an arsenic content up to 6.9%, as a chisel from Umm an-Nar (Berthoud 1979: 32, tab. 5). High arsenic content was also observed in previous analyses carried out on finds from HD-6 (see Chapter 6).

Nickel (Ni)

Nickel was observed in all the finds with variable concentrations up to 4% (Figures 12.12-12.15). The earliest finds generally have a Ni content lower than other items. Only a few items have a Ni content under the MDL. An uncommon copper-nickel-arsenic-tin alloy was detected in a fishhook found at GAS-1, on the surface; it was characterized by a high Ni content, around 7%. As for arsenic, two different groups were defined depending on Ni content. Generally, higher values of arsenic match with higher values of nickel, because there is a correlation between the two elements (Figure 12.16). A similar correlation also occurs between nickel and cobalt (Figure 12.17). The relative diagrams show that both arsenic and nickel have the lowest average value in the finds from HD-6 than in items from RJ-2, even if there are two groups with a similar concentration, as highlighted in the table (Table 12.7).

Lead (Pb)

Lead was found sporadically, in trace amounts, in the objects from RJ-1 and HD-6, and it is very rare in the earliest finds, where the majority of finds has no lead at all (Figure 12.2). It occurs more in artefacts from RJ-2, were lead has an average value of 0.5% (fig. 18). Less than ten objects from that site have a lead content of around 1%; in these items lead is associated with a higher arsenic content. The highest Pb content (2.10%) was observed in a chisel from RJ-2 (DA 8687). Nevertheless, this value is too low to suppose a deliberate addition. Lead reaches 17% in a blocklet from RJ-2 (DA 12735), where it is associated with 18% As.

Antimony (Sb)

Antimony has an average concentration in the order of 0.5-0.7%, just above the MDL (Figure 12.19). The finds from RJ-2 have more frequently an antimony content higher than the MDL compared with the items from other sites. Sb was detected in 25% of the objects from RJ-2, while only in 10% of the finds from HD-6. The highest antimony content was detected in an awl from HD-6 (DA 2247), 4.06%; this value is associated with an anomalous silver content (2.44% Ag). Other high Sb contents were observed in a blocklet from RJ-2, 3.04%, and in a shell-opener from the same site, 1.04%. Sb value approaches 1% in some finds from RJ-2; it is associated with an arsenic content between 2% and 3%: As and Sb are in fact strongly correlated.

Tin (Sn)

Tin is not a common element in the set of analyzed materials (Figure 12.20). Only one fishhook from GAS-1 contains some tin, around 1%: it was made of an unusual copper-nickel-arsenic-in alloy, where Ni is around 7%. Nickel was detected more frequently in the artefacts from RJ-2: 11 objects in comparison with only two items from HD-6. Furthermore, tin is below 1% in both the HD-6 finds (DA 2261: 0.56%; DA 2278: 0.23%): a clear indication that the metal was an unintentional alloy. Also, some of the RJ-2 finds have a low Sn content, below 1%, with an average concentration value of 0.5%. These low tin items are probably connected with the recycling of bronze objects; these objects, probably imported from other areas, were therefore melted together with Omani copper. The recycling practice was also hypothesized for other contexts of Southwestern Arabia belonging to the Umm an-Nar period (Weeks 2003: 94-95). Only seven objects have concentration values higher than 1%: two awls (DA 8367: 1.57%; 12741: 2.40%), two razors (DA 12732: 2.42%; DA 12733: 16.4%), a blocklet (DA 12738: 1.11%), a fishhook (DA 12892: 1.33%) and a ring (DA 8697: 1.31%). According to these results, tin bronze alloy was mainly used to make ornaments (Figures 12.21-12.23). It should be noted that in most of the bronze items tin was associated with greater contents of arsenic and nickel. A razor had a very high tin content: 16.4%. Probably this value could be connected to corrosion processes that produced an enrichment of tin on the surface because of the leaching of copper.

Table 7 represents average values for elementary concentration, calculated for the groups of finds distinguished for provenance and chronology (RH-10, GAS-1, HD-6, RJ-2, RJ-1). The average was evaluated on all artefacts belonging to each group (see first row, next to the symbol of the element). However, standard deviation was very high: the sets are in fact irregular, because "normal" objects with similar values are associated with "exceptional" finds that have abnormal values for a given element. We pointed out these "abnormal" objects and we recalculated average without them, with standard deviation. For some of the elements, subgroups of objects were identified, characterized by similar average values. The table indicates not only the average values, but also the number of objects on which the average was calculated.

Analyses of the crucibles

Most of the crucibles were examined on both sides, inside and outside, in order to detect possible differences in the concentration of metallic elements (Table 12.8). In fact, we have to expect a higher content of copper in the inner surface of a crucible, because it is devoted to contain molten copper. The clay matrix is very similar in concentration for all analyzed samples. Only for items B2497_30 and B2505_33 we found significant traces of copper and lead; both pieces come from Ras Al-Jinz RJ-2. It should be stressed that the presence of detected metals confirms the nature of the potsherd as a crucible; nevertheless, the lack of this evidence does not totally exclude this possibility. In fact, copper or other metals could not be fixed on the clay surface for many different reasons, and therefore they were not detected by XRF.

Analyses of a copper based prehistoric cosmetic

The remains of a prehistoric cosmetic were found at Ras Al-Hadd HD-5, inside the valve of a cockleshell. The cosmetic was recovered in an Umm an-Nar context that dates back to the second half of the 3rd millennium BC. It had the aspect of a green, metallic powder. A small sample of this powder was analyzed by XRF (Table 12.9). The main component was copper sulfide (79% Cu; 1.1% S), to which some lime (calcium oxide, CaO) was added in order to produce a plastic, malleable amalgam.

Table 12.1. Metal finds from Wadi Shab GAS-1. Alloy normalized concentration
MDL (Minimum Detection Limit) and MRE (Minimum Relative Error) are specified for the elements.

Finds (DA Nos.)		% Ag	% As	% Co	% Cu	% Fe	% Ni	% Pb	% Sb	% Sn	% Zn
	MDL	0.25	0.05	0.03	0.02	0.02	0.05	0.30	0.40	0.20	0.05
	MRE	8.00	5.00	5.00	0.40	6.00	3.00	10.00	20.00	8.00	7.00
GAS1_DA32400_1	unfin. tool		0.19	0.02	98.7	0.33	0.61				0.13
GAS1_DA32400_2	awl			0.03	99.3	0.11	0.06				0.11
GAS1_DA32400_3	pin	5.09	0.61		93.6	0.12	0.54				
GAS1_DA32400_4	awl	0.93	0.60		97.7	0.07			0.51		0.17
GAS1_DA32400_5	hook	2.46	1.53		86.0	0.68	7.22		0.92	1.22	
GAS1_US101_001	unfin. tool		0.71		98.9	0.06	0.10				0.15
GAS1_US101_002	awl		0.77		98.4	0.32	0.24				0.16
GAS1_US101_003	awl		1.16		98.3	0.14	0.24				0.15
GAS1_US100_004	chisel		2.19	0.39	95.1	0.07	1.85				0.10
GAS1_US100_005	chisel		0.86	0.02	97.5	0.21	0.87		0.42		0.12
GAS1_US100_006	unfin. or netting tool	0.27	1.41		97.4	0.28	0.07		0.44		0.14
GAS1_US100_007	drill point	8.60	3.39	0.06	83.8	1.10	2.99				
GAS1_US100_008	pin	1.24	0.16	0.02	98.1	0.23	0.11				0.13
GAS1_US100_009	chisel			0.03	99.4	0.17	0.15				0.15
GAS1_US101_011	awl		1.37		98.0	0.45	0.07				0.14
GAS1_US101_012	awl	1.84	0.36	0.03	97.1	0.29	0.24				0.11
GAS1_US101_013	awl	3.53	0.62	0.03	93.3	0.20	1.08		1.17		0.10
GAS1_US101_014	chisel		1.43	0.02	97.8	0.35	0.05				0.14
GAS1_US101_015	hook	1.49	0.19		97.9	0.18	0.09				0.14
GAS1_US101_016	pin	0.64	0.16	0.03	98.5	0.06	0.06		0.42		0.15
GAS1_US101_017	drill point	1.83	0.17		97.4	0.21	0.28				0.13
GAS1_US101_018	unfin. or netting tool	0.56	0.21	0.02	98.8	0.12	0.16				0.11
GAS1_US101_019	pin	0.70	0.27		97.9	0.48	0.50				0.14
GAS1_US101_020	awl	0.41	0.24	0.02	98.7	0.09	0.06		0.41		0.08
GAS1_US100_021	drill point	1.51	0.33	0.03	97.3	0.24	0.41				0.14
GAS1_US100_022	awl	1.97		0.02	96.4	0.11	0.22	0.44	0.72		0.10
GAS1_US100_023	awl	0.41		0.02	99.1	0.11		0.25			0.13
GAS1_US100_024	awl	0.63	1.31	0.03	95.4	0.07	2.46				0.13
GAS1_US100_025	chisel		0.20	0.02	99.2	0.15	0.33				0.12
GAS1_US100_026	chisel	0.77	0.31	0.02	98.3	0.24	0.19				0.13
GAS1_US100_027	awl	1.18	0.90	0.12	96.0	0.12	1.01		0.53		0.12
GAS1_US100_028	awl		0.40	0.02	99.0	0.12	0.22				0.13
GAS1_US100_029	awl	1.40	0.88	0.02	96.3	0.22	1.03				0.13
GAS1_US100_031	hook	2.11		0.03	97.3	0.20	0.18				0.16
GAS1_US100_032	unfin. tool		0.63	0.02	99.1	0.14					0.13
GAS1_US100_033	chisel	1.25			98.2		0.20				0.14
GAS1_US100_034	unfin. tool	0.38	0.51	0.02	98.0	0.18	0.16		0.42		0.16
GAS1_US100_035	awl	0.97		0.02	98.8	0.13					0.12

Finds (DA Nos.)		% Ag	% As	% Co	% Cu	% Fe	% Ni	% Pb	% Sb	% Sn	% Zn
GAS1_US100_036	awl	0.28	0.93	0.02	98.2	0.21	0.05				0.14
GAS1_US100_037	unfin. tool	0.32	0.46	0.02	96.0	0.38	0.24	2.40			0.15
GAS1_US100_038	chisel		0.23	0.03	98.5	0.50	0.21				0.16
GAS1_US100_039	awl	0.94	0.68		97.8	0.24	0.15				0.18
GAS1_US100_040	awl	0.49	3.39	0.04	92.5	0.21	2.26	1.03			0.09
GAS1_US100_041	hook	1.07	0.22	0.02	98.1	0.14	0.36				0.14
GAS1_US100_042	awl	0.59	0.26	0.03	98.0	0.19	0.51				0.11
GAS1_US100_043	awl		0.19	0.02	99.5	0.06					0.12
GAS1_US100_044	pin	1.15	0.33		96.4	0.42	1.64				0.10
GAS1_US100_045	chisel	0.56	0.28	0.02	98.7	0.15	0.12				0.15
GAS1_US100_046	awl	1.23	0.12	0.02	98.4	0.09					0.14
GAS1_US100_047	chisel	0.54	0.30	0.03	98.3	0.15	0.17			0.42	0.12
GAS1_US100_048	awl		0.65	0.02	98.2	0.51	0.45				0.12
GAS1_US100_049	pin	0.28	0.33	0.02	98.8	0.20	0.23				0.11
GAS1_US100_050	awl		0.20	0.02	99.5	0.11					0.16
GAS1_US100_051	pin	0.55	0.25	0.03	98.5	0.31	0.19				0.14
GAS1_DA15238	hook	1.68	0.19	0.02	97.7	0.10	0.16				0.14
GAS1_DA15240	netting tool	0.73	0.09	0.02	98.9	0.05	0.17				

Table 12.2. Metal finds from Ras Al-Hamra RH-10. Alloy normalized concentration. MDL (Minimum Detection Limit) and MRE (Minimum Relative Error) are specified for the elements.

Finds (DA Nos.)		% Ag	% As	% Co	% Cu	% Fe	% Ni	% Pb	% Sb	% Sn	% Zn
	MDL	0.25	0.05	0.03	0.02	0.02	0.05	0.30	0.40	0.20	0.05
	MRE	8.00	5.00	5.00	0.40	6.00	3.00	10.00	20.00	8.00	7.00
RH10_2639_02	pointed tool	1.05			98.1	0.27	0.16				0.14
RH10_2639_04	pointed tool	0.99	0.33		97.1	1.30	0.16				0.11
RH10_2639_06	awl	0.87	1.10	0.06	95.5	0.19	0.99	0.63	0.53		0.11
RH10_2639_09	hook		0.84		98.0	0.64	0.19				0.12
RH10_2639_10slag	copper fragment		0.63	0.13	94.7	1.54	2.31				0.13
RH10_2639_11	awl	0.36	3.50	0.05	92.5	0.69	2.10	0.70			0.08
RH10_2639_12	drill point	2.09	0.88	0.06	96.3	0.51	0.16				
RH10_2639_14	pointed tool		1.72	0.03	95.6	0.66	1.14		0.45		0.10
RH10_2639_15_1	awl	0.28	1.11	0.05	96.5	0.10	0.80	1.04			0.09
RH10_2639_15_2	pin	1.58	0.39	0.06	96.4	0.40	0.51	0.63			0.08
RH10_2639_17	pointed tool	0.59	5.80		92.8	0.49	0.16				0.15
RH10_2639_18	pointed tool	0.65	3.99		92.6	1.25	1.14	0.30			0.12
RH10_2639_19	awl	1.45	1.23		95.8	0.43	0.21		0.78		0.10

Chemical-Physical Analyses by Energy Dispersive X-Ray Fluorescence (EDXRF)

RH10_2639_20	pointed tool		1.56	0.06	96.9	0.71	0.39			0.12
RH10_2639_29	hook	1.49	1.53		95.3	0.24	1.37			0.10
RH10_2639_30	hook	1.16	0.72		96.5	0.41	0.77	0.34		0.12
RH10_2639_31	hook	1.64	2.16		88.8	0.14	2.92	4.32		
RH10_6587	hook	1.52	1.57	0.07	94.6	0.27	1.87			0.12
RH10_6595	awl	1.83	0.76	0.03	94.9	0.08	0.70	0.91	0.70	0.12
RH10_10979	chisel	6.26	0.42		92.5	0.36	0.50			
RH10_6596	hook	2.01	0.17	0.03	96.9	0.82				0.10
RH10_6973_35a	awl	0.52	0.46	0.03	97.3	1.36	0.19			0.15
RH10_6973_35b	awl		2.43	0.04	94.0	1.28	0.90	0.34	0.65	0.13

Table 12.3. Metal finds from Ras Al-Hadd HD-6. Alloy normalized concentration.
MDL (Minimum Detection Limit) and MRE (Minimum Relative Error) are specified for the elements.

Finds (DA Nos.)		% Ag	% As	% Co	% Cu	% Fe	% Ni	% Pb	% Sb	% Sn	% Zn
	MDL	*0.15*	*0.05*	*0.03*	*0.03*	*0.03*	*0.05*	*0.12*	*0.50*	*0.15*	*0.06*
	MRE	*12.0*	*6.0*	*6.0*	*0.5*	*0.5*	*4.0*	*15.0*	*15.0*	*6.0*	*12.0*
HD6_2247	awl	2.44	0.91		91.6	0.17	0.77		4.06		
HD6_2249	chisel		0.47	0.06	96.3	1.76	1.15				0.25
HD6_2261	hook	0.34	0.17		96.8	0.92	0.43		0.56	0.56	0.17
HD6_2262	awl	0.40	1.23		94.8	1.02	2.43				0.15
HD6_2263	chisel	0.15	0.14	0.09	98.9	0.32	0.20				0.17
HD6_2276	blade	0.44	0.10		99.1	0.05	0.16				0.16
HD6_2277	hook	0.44	0.31		96.7	0.93	0.79	0.18	0.48		0.15
HD6_2278	hook	0.28	0.15		97.5	1.31	0.11	0.14		0.23	0.23
HD6_2279	hook	0.86			94.9	2.54	0.25	0.47	0.70		0.30
HD6_2367	pin	0.30	0.19		98.8	0.03	0.45				0.19
HD6_11554	blocklet		0.21	0.03	98.8	0.40	0.38				0.15
HD6_11578	hook		0.10		99.3	0.15	0.29				0.13
HD6_11585	pin	0.38	1.91		95.4	0.07	2.12				0.13
HD6_11586	blocklet		1.46	0.02	98.1	0.05	0.25				0.14
HD6_11590	blocklet fragment		0.41	0.04	98.5		0.86				0.15
HD6_12484	hook		0.25	0.03	99.3	0.03	0.28				0.13
HD6_14071	blocklet	0.29	1.91		95.7	0.18	1.94				
HD6_14702	hook		0.54		98.7	0.18	0.44				0.16
HD6_14707_01	awl		2.44	0.04	94.1	0.07	3.25				0.12
HD6_14707_02	unfin. tool		0.29		99.2	0.10	0.22				0.16
HD6_14724	awl		0.26	0.03	99.1	0.16	0.32				0.18
HD6_14981	hook		0.55		99.1		0.12				0.22
HD6_15072	chisel		0.20		99.3	0.06	0.25				0.15
HD6_15073	awl		0.18	0.02	99.6	0.03					0.15
HD6_15076	awl		0.47		98.0	1.31	0.08				0.19

HD6_15078	chisel	0.57	0.12	0.04	99.2	0.05	0.05				
HD6_15082	dagger		3.31		93.8	0.05	2.71				0.10
HD6_15088	crochet		0.18		99.1	0.23	0.21				0.22
HD6__15091	crochet		0.14		97.8	1.11	0.66				0.27
HD6_15092	crochet			0.04	98.5	1.09	0.13				0.28
HD6_15667_01	hook		0.41		97.8	1.63					0.18
HD6_15667_02	hook		2.28		97.3	0.03	0.17				0.22
HD6_17270	awl		0.24	0.04	98.8	0.03	0.91				
HD6_17271	chisel	0.44	0.06	0.03	99.2	0.05	0.09				0.14
HD6_17756	dagger		4.19		90.8	0.24	4.60				0.13
HD6_17805	unfin. tool		3.47		92.1	0.08	3.57	0.61			0.14

Table 12.4. Metal finds from Ras Al-Hadd HD-10. Alloy normalized concentration.
MDL (Minimum Detection Limit) and MRE (Minimum Relative Error) are specified for the elements.

Finds (DA Nos.)		% Ag	% As	% Co	% Cu	% Fe	% Ni	% Pb	% Sb	% Sn	% Zn
	MDL	*0.25*	*0.05*	*0.03*	*0.02*	*0.02*	*0.05*	*0.30*	*0.40*	*0.20*	*0.05*
	MRE	*8.00*	*5.00*	*5.00*	*0.40*	*6.00*	*3.00*	*10.00*	*20.00*	*8.00*	*7.00*
HD10_14334	knife	0.31	0.88	0.02	96.5	0.10	2.02				0.14
HD10_14334	rivet		0.75		93.0	5.60	0.32				0.29
HD10_14340	awl	2.48			95.6	0.40	0.21		1.34		
HD10_3_2_L3	fragment	0.94	0.25		97.2	1.17	0.28				0.13
HD10_4_2_L101	fragment	2.85	0.33		95.5	0.63	0.68				
HD10_4_2_L1	fragment	3.12			92.3	4.43	0.16				

Table 12.5. Metal finds from Ras Al-Jinz RJ-2. Alloy normalized concentration.
MDL (Minimum Detection Limit) and MRE (Minimum Relative Error) are specified for the elements.

Finds (DA Nos.)		% Ag	% As	% Co	% Cu	% Fe	% Ni	% Pb	% Sb	% Sn	% Zn
	MDL	*0.15*	*0.05*	*0.03*	*0.03*	*0.03*	*0.05*	*0.12*	*0.50*	*0.15*	*0.06*
	MRE	*12.00*	*6.00*	*6.00*	*0.50*	*0.50*	*4.00*	*15.00*	*15.00*	*6.00*	*12.00*
RJ2_1382	flat axe		0.70	0.05	98.5	0.17	0.29	0.13			0.14
RJ2_8335	hook?	0.21	0.07		99.1	0.42	0.07				0.16
RJ2_8336	hook		0.54		98.7	0.28	0.33				0.18
RJ2_8366	blocklet		0.18		98.4	0.22	1.03				0.12
RJ2_8367	awl		3.34	0.07	90.3	0.39	3.13	1.04		1.57	0.16
RJ2_8371	shell-opener	0.28	3.76		90.3	0.34	2.50	1.67	1.04		0.13
RJ2_8374	blade		2.53		94.8	0.48	1.11	0.23	0.66		0.17
RJ2_8477	blocklet		0.37	0.02	97.0	0.30	1.12	0.16	0.80		0.17
RJ2_8506	flat ingot		1.31		97.5	0.24	0.77				0.21
RJ2_8574	small knife		1.74		95.9	0.29	1.12	0.76			0.19
RJ2_8577	awl		2.70	0.03	93.8	0.47	2.50	0.40			0.13

RJ2_8687	chisel		2.10	0.02	94.6	0.07	0.96	2.10		0.11
RJ2_8697	ring	0.91	1.31		94.1	0.23	0.79	1.21	1.31	0.18
RJ2_8703	unfin. hook?		0.64	0.04	97.8	0.49	0.67	0.22		0.19
RJ2_8707	hook		0.37		99.0	0.54	0.06			
R12_10012	hook		0.51	0.08	98.8	0.07	0.43			0.14
RJ2_10032	shell-opener		1.84	0.03	97.5	0.03	0.39			0.16
RJ2_10038	hook	0.28	1.07	0.04	96.3	0.46	0.91	0.30	0.43	0.17
RJ2_10043	chisel	0.22	0.33	0.06	98.3	0.13	0.80			0.15
RJ2_10055	blocklet	0.25	0.45	0.03	97.2	0.11	1.00	0.31	0.50	0.15
RJ2_10056	drill point	0.31	0.32		98.7	0.30	0.06			0.27
R12_10057	unfin. hook		0.86		98.1	0.54	0.30			0.18
RJ2_10062	chisel		0.93	0.19	95.5	0.52	1.76	0.35	0.73	
RJ2_10064	flat axe	0.16	1.80	0.04	95.1	0.57	1.17	0.39	0.60	0.13
R12 10186	chisel		0.28	0.03	98.7	0.05	0.17	0.18	0.40	0.15
RJ2_10853	awl		1.20	0.04	98.5	0.16				0.07
RJ2_10949	blocklet		1.80	0.03	95.2	0.05	1.06	1.20	0.60	0.11
R12_10959	iron hook?				100					
RJ2_11997	blocklet	0.27	4.84	0.03	90.7	0.23	2.89		0.93	0.14
RJ2_11999	unfin. hook		4.25	0.08	92.3	0.71	1.40	1.06		0.17
RJ2_12076	shell-opener	0.26	2.29		94.6	0.15	0.85	0.26	0.80	0.57	0.17
RJ2_12086	casting residue	0.35	0.18	0.02	98.9	0.03	0.26	0.11		0.17
RJ2_12089	crochet	0.19	2.47	0.04	96.0	0.69	0.33	0.12		0.17
RJ2_12357	small knife	0.19	1.44	0.08	95.9	0.26	1.88	0.13		0.13
RJ2_12474	hook		1.64		95.7	0.34	1.89	0.26		0.13
RJ2_12479	unfin. hook	0.43	1.33		96.8	0.22	0.66	0.42		0.16
RJ2_12487	blocklet		1.61	0.04	95.9	0.04	1.14	0.80	0.40	0.10
RJ2_12488	blocklet	0.15	1.02	0.04	96.9	0.33	0.32	0.60	0.50	0.17
R12_12489	blocklet	0.26	2.20	0.03	93.7	0.13	1.54	1.10	0.90	0.12
RJ2_12490	blocklet	0.19	2.50	0.07	95.8	0.33	0.74	0.24		0.14
RJ2_12492	blocklet	0.24	0.55	0.04	97.7	0.32	0.38	0.60		0.18
RJ2_12493	blocklet		0.49		98.9	0.16	0.22			0.20
RJ2_12497	pin	0.21	0.43	0.02	98.8	0.16	0.17			0.20
RJ2_12503	shell-opener		1.53	0.05	95.1	0.37	1.20	0.23	0.64	0.64	0.20
R12_12504	crochet		2.52	0.06	94.2	0.33	1.82	0.88		0.14
RJ2_12507	flat ingot		2.40	0.04	96.8	0.11	0.34	0.14		0.16
RJ2_12508	blade		1.51		95.1	0.72	0.84	1.11	0.60	0.16
RJ2_12548	chisel	0.25	3.64	0.04	92.8	0.68	1.98	0.42		0.15
RJ2_12673	casting residue		0.14		99.4	0.20	0.11			0.12
RJ2_12732	razor		0.89		95.3	0.60	0.16	0.50	2.42	0.15
RJ2_12733	razor	0.26	1.95		79.1	0.58	1.19	0.35	16.4	0.12

Finds (DA Nos.)		% Ag	% As	% Co	% Cu	% Fe	% Ni	% Pb	% Sb	% Sn	% Zn
RJ2_12734	chisel	0.28	2.60	0.04	95.1	0.16	0.79		0.83		0.17
RJ2_12735	blocklet	0.81	18.2	0.07	57.9		2.63	17.2	3.04		0.09
RJ2_12736	razor		2.07		96.1	0.49	0.68	0.52			0.14
RJ2_12738	blocklet	0.23	1.41	0.04	95.3	0.29	1.22	0.30		1.11	0.11
RJ2_12739	chisel	0.33	1.25		97.1	0.26	0.71	0.21			0.15
RJ2_12741	awl	0.32	1.42		94.2	0.48	0.72	0.29		2.40	0.16
RJ2_12742	chisel		2.63	0.06	95.5	0.17	0.76	0.22	0.51		0.14
RJ2_12750	ring	0.35	0.18		97.7	0.29	0.52	0.74			0.20
RJ2_12774	hook	0.29	0.71		97.3	0.11	0.51	0.90			0.13
RJ2_12775	ring		2.26	0.03	96.4	0.13	0.80	0.18			0.18
RJ2_12785	hook	0.19	0.96		97.6	0.09	0.49	0.70			
RJ2_12786	pin		0.17	0.05	98.9	0.27	0.42				0.17
RJ2_12791	ring		0.94	0.04	97.2	0.31	0.94	0.41			0.16
RJ2_12792	hook (Harappan)		0.28	0.03	98.0	1.25	0.18				0.28
RJ2_12896	blocklet	0.21	0.34	0.04	98.9	0.14	0.14	0.12			0.14
RJ2_12897	hook?	0.29	1.33	0.05	95.5	0.25	0.62	0.51		1.33	0.14
RJ2_12906	awl		0.62	0.04	97.3	0.30	1.02		0.52		0.18
RJ2_12913	shell-opener		0.86		97.6	0.33	0.72	0.43			

Table 12.6. Metal finds from Ras Al-Hadd HD-5, Ras Al-Jinz RJ-1, RJ-4 and RJ-6. Alloy normalized concentration. MDL (Minimum Detection Limit) and MRE (Minimum Relative Error) are specified for the elements.

Finds (DA Nos.)		% Ag	% As	% Co	% Cu	% Fe	% Ni	% Pb	% Sb	% Sn	% Zn
	MDL	*0.15*	*0.05*	*0.03*	*0.03*	*0.03*	*0.05*	*0.12*	*0.50*	*0.15*	*0.06*
	MRE	*12.00*	*6.00*	*6.00*	*0.50*	*0.50*	*4.00*	*15.00*	*15.00*	*6.00*	*12.00*
RJ1_8681	awl		0,04%		99,6%	0,05%	0,14%				0,12%
RJ1_14404	flat chisel		2,10%	0,04%	96,7%	0,12%	0,94%				0,08%
RJ1_14405	casting residue		0,08%	0,16%	97,4%	1,42%	0,87%				0,12%
RJ1_14406	hook		0,36%	0,19%	96,9%	0,31%	2,11%				0,19%
RJ1_14313	needle		1,47%	0,25%	95,5%	0,40%	2,21%				0,19%
RJ1_14314	awl		0,79%		97,8%	0,12%	0,93%	0,22%			0,11%
RJ4_8696	small knife		2,55%	0,12%	90,9%	0,48%	5,93%				
HD5_12803	seal		0,96%	0,03%	98,2%	0,34%	0,35%				0,15%
RJ6_8621	pin	0,30%	1,72%	0,03%	96,5%	0,02%	1,29%				0,14%

Table 12.7. Average for element in objects from Wadi Shab GAS-1, Ras Al-Hamra RH-10, Ras Al-Hadd HD-6 and HD-10, Ras Al-Jinz RJ-2 and RJ-1, analyzed using XRF

	RJ-1 (6)		HD-6 (36)		RJ-2 (69)		HD-10 (6)		RH-10 (23)		GAS-1 (56)	
	(%)	No.	(%)	No.	(%)	No.	(%)	No.	(%)	No.	(%)	No.
Ag		0	0,56 ± 0,59	13	0,30 ± 0,16	31	1,9 ± 1,2	5	1,5 ± 1,3	18	1,4 ± 1,5	39
			2,44	1	0,86 ± 0,07	2	3,0 ± 0,2	2	6,3	1	8,6	1
			0,41 ± 0,18	12	0,26 ± 0,06	29	2,5	1	2,05 ± 0,04	2	4,3 ± 1,1	2
							0,6 ± 0,4	2	1,59 ± 0,14	6	2,0 ± 0,3	6
									1,02 ± 0,12	4	1,2 ± 0,2	12
									0,48 ± 0,16	5	0,51 ± 0,16	18
As	0,81 ± 0,83	6	0,86 ± 1,11	34	1,65 ± 2,28	69	0,55 ± 0,31	4	1,5 ± 1,4	22	0,7 ± 0,7	49
	1,78 ± 0,44	2	4,19	1	18,20	1	0,81 ± 0,09	2	5,8	1	3,4 ± 0,1	2
	0,32 ± 0,35	4	2,47 ± 0,99	9	2,18 ± 0,88	37	0,29 ± 0,06	2	3,75 ± 0,35	2	1,48 ± 0,33	7
			0,28 ± 0,19	24	0,54 ± 0,30	31			1,60 ± 0,46	9	0,75 ± 0,13	12
									0,56 ± 0,21	10	0,27 0,10	28
Co	0,16 ± 0,09	6	0,04 ± 0,02	13	0,05 ± 0,03	43	0,02	1	0,05 ± 0,03	13	0,04 ± 0,06	42
									0,13	1	0,39	1
									0,05 ± 0,02	12	0,12	1
											0,02 ± 0,01	
Cu	97,3 ± 1,4	6	97,3 ± 2,4	36	95,6 ± 5,5	69	95,0 ± 1,9	6	95,2 ± 2,2	23	97,3 ± 2,8	56
					96,4 ± 2,2	7	97,2	1	98,03 ± 0,09	2	98,4 ± 0,6	44
					79,1	1	95,9 ± 0,6	3	95,9 ± 1,0	16	96,0 ± 0,5	7
					57,9	1	92,7 ± 0,5	2	92,6 ± 0,2	4	93,1 ± 0,6	3
									88,8	1	84,9 ± 1,5	2
Fe	0,40 ± 0,52	6	0,48 ± 0,64	34	0,31 ± 0,21	69	2,1 ± 2,5	6	0,6 ± 0,4	23	0,22 ± 0,18	55
	1,42	1	2,54	1	1,25	1	5,60	1	1,35 ± 0,12	5	1,10	1
			1,23 ± 0,30	9			4,43	1	0,70 ± 0,07	5	0,68	1
	0,20 ± 0,15	5	0,12 ± 0,10	24	0,30 ± 0,18	68	1,17	1	0,3 ± 0,14	13	0,44 ± 0,06	7
							0,38 ± 0,25	3			0,23 ± 0,05	21
											0,11 ± 0,03	25
Ni	1,20 ± 0,80	6	0,90 ± 1,16	34	0,90 ± 0,71	68	0,6 ± 0,7	6	0,9 ± 0,8	22	0,63 ± 1,17	49
	2,16 ± 0,07	4	4,60	1	2,16 ± 0,55	12	2,02	1	2,92	1	7,2	1
	0,91 ± 0,04	2	2,67 ± 0,64	6	0,63 ± 0,36	56	0,68	1	2,09 ± 0,22	3	2,6 ± 0,4	3
	0,14	1	0,85 ± 0,17	6			0,24 ± 0,07	4	0,98 ± 0,23	8	1,3 0,4	6
			0,23 ± 0,12	21					0,26 ± 0,14	10	0,22 ± 0,115	39
Pb	0,22	1	0,35 ± 0,23	4	0,87 ± 2,42	49		0	1,0 ± 1,3	9	1,0 ± 0,9	4
					17,20	1			4,32	1	2,4	1
					1,31 ± 0,38	8			1,0 ± 0,1	2	1,03	1
					0,38 ± 0,22	40			0,5 ± 0,2	6	0,35 ± 0,13	2
Sb		0	1,45 ± 1,74	4	0,80 ± 0,59	18	1,34	1	0,62 ± 0,13	5	0,58 ± 0,25	11
			4,06	1	3,04	1					0,93 ± 0,23	3
			0,58 ± 0,11	2	0,67 ± 0,19	17					0,45 ± 0,04	8
Sn		0	0,40 ± 0,23	2	2,62 ± 4,63	11		0		0	1,22	1
					16,40	1					±	
					1,69 ± 0,58	6					±	
					0,56 ± 0,09	4					±	
Zn	0,14 ± 0,04	6	0,18 ± 0,05	32	0,16 ± 0,03	65	0,19 ± 0,09	3	1,12 ± 0,02	20	0,13 ± 0,02	52

Table 12.8. Crucibles from Ra's Al-Jinz RJ-2; int.: inside, ext.: outside. Majority elements (maj): + = 1.00%; ++ = 5%; +++ = 15%; Minority elements (min): + = 0.50%; ++ = 5%; +++ = 10%; ++++ = 15%. Trace elements (trace): + = 0.20%; ++ = 1%; +++ = 5%; ++++ = 10%. MDL (Minimum Detection Limit) and MRE (Minimum Relative Error) are specified for the components.

Finds (DA Nos.)	Ag	Al₂O₃	CaO	Cl	CuO	Fe₂O₃	K₂O	MnO	Ni	Pb	SiO₂	SO₃	Sr	TiO₂
MDL	0.15%	1.50%	0.12%	0.15%	0.03%	0.03%	0.15%	0.05%	0.05%	0.12%	0.30%	0.60%	0.05%	0.06%
MRE	11.00%	15.00%	1.50%	5.00%	1.00%	1.00%	4.00%	15.00%	15.00%	0.50%	10.00%	5.00%	2.00%	5.00%
	min	maj	maj	min	trace	maj	min	trace	trace	trace	maj	trace	min	min
RJ2_B2497_30_ext		++	++++	+	+	+++		+	tr	++	+++	+	+	+
RJ2_13249730_int_b			++	+	++++	+				++++	+		tr	
RJ2_13249730_int_v			+++	+	++++				++++	++++	+		tr	tr
RJ2_B2497_31_int		++	+++	tr	+	+++	++	+	tr	+	+++		+	+
RJ2_B2505_32		++	+++		tr	+++	++	tr	tr	+	+++		+	+
RJ2_B2505_33		++	+++	+	+	+++	++	tr	tr	++++	++		tr	+
RJ2_B2505_34	+	+	+++		+	+++	+	+	tr	+	+++		+	+
RJ2_B2505_35		++	++++		+	+++	+	+	tr		++	++++	+	+
RJ2_B2505_36_ext	+		++++		+	+++	+	+	tr		+++	++	+	+
RJ2_B2505_36_int	+		++++		+	+++	+	+	tr		+++		+	+
RJ2_B2505_37	+		+++	+	+	++	++	+	tr	+	+++		+	+

Table 12.9. Metallic cosmetic amalgam inside a cockle shell from Ras Al-Hadd HD-5. MDL (Minimum Detection Limit) and MRE (Minimum Relative Error) are specified for the elements.

Finds (DA Nos.)	%Ag	%As	%Ca	%Cl	%Co	%Cr	%Cu	%Fe	%K	%Ni	%Pb	%S	%Se	%Sr	%Zn
MDL	0.15	0.05	0.12	0.15	0.03	0.05	0.03	0.03	0.14	0.05	0.12	0.30	0.10	0.05	0.06
MRE	12.00	6.00	2.00	5.00	6.00	15.00	0.50	0.50	4.00	15.00	15.00	5.00	7.00	5.00	12.00
HD5_11760	-	0.09	7.30	4.10	0.16	-	7.21	12.50	-	0.26	-	1.10	-	0.78	0.07

Chemical-Physical Analyses by Energy Dispersive X-Ray Fluorescence (EDXRF)

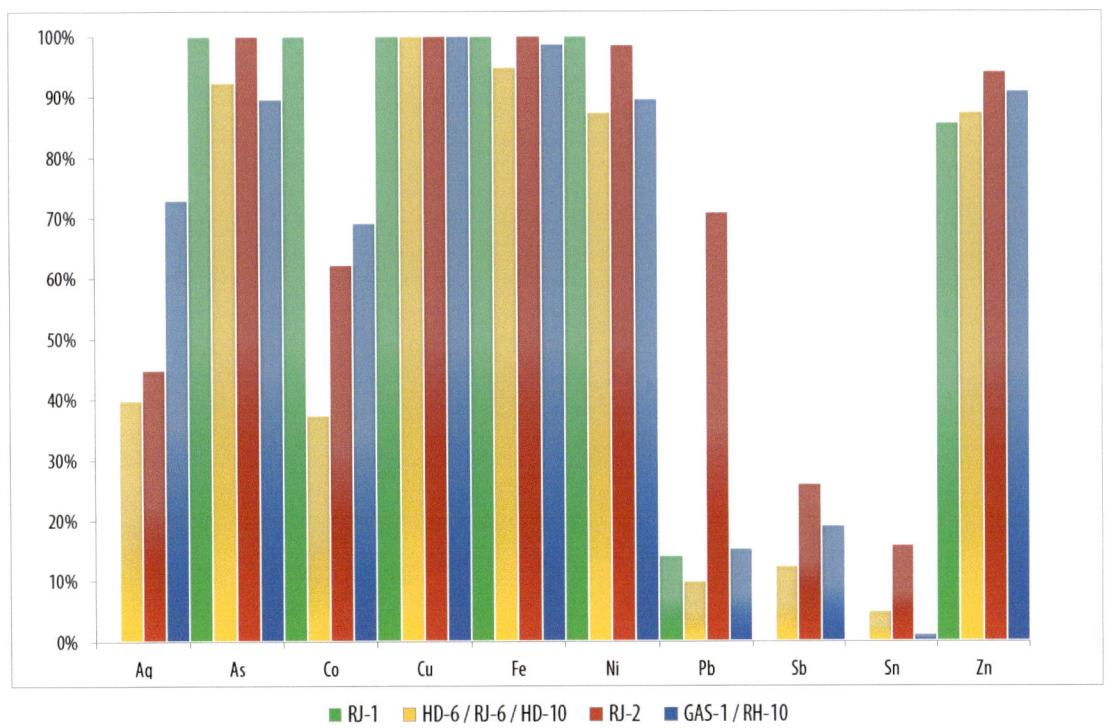

Figure 12.2. Histogram of presence frequency of the analyzed elements in objects from Wadi Shab GAS-1, Ras Al-Hamra RH-10, Ras Al-Hadd HD-6, Ras Al-Jinz RJ-2 and Ras Al-Jinz RJ-1.

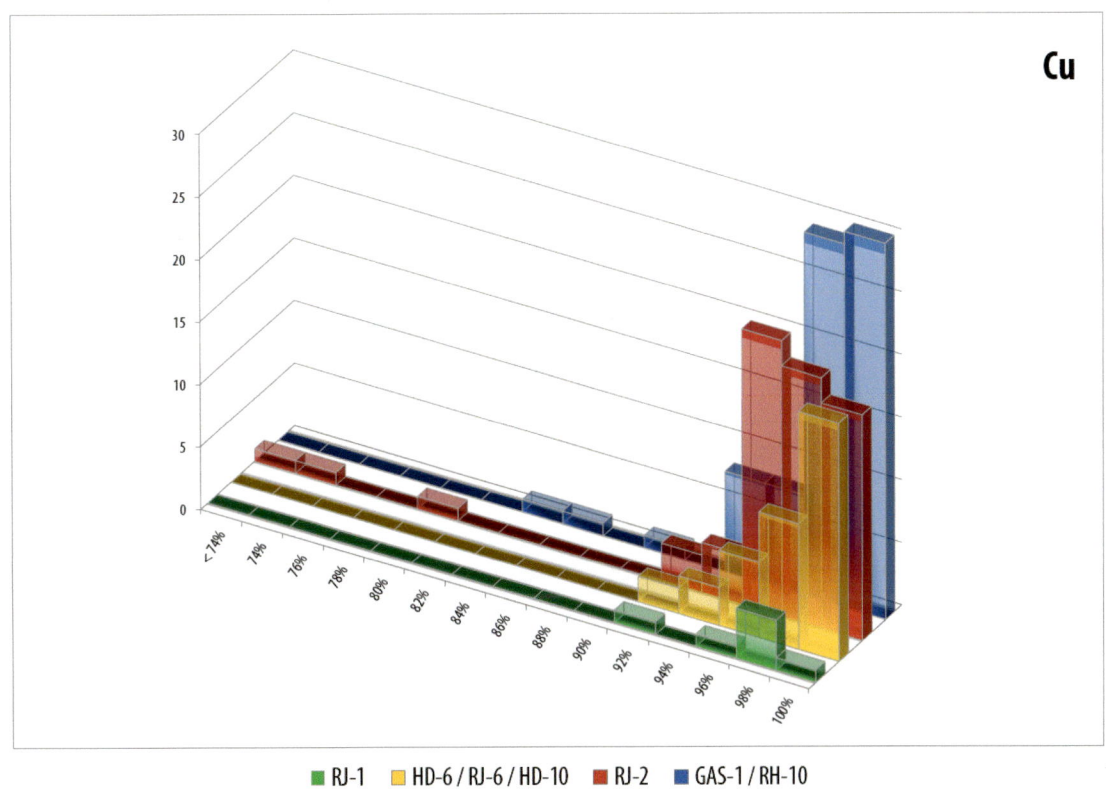

Figure 12.3. Copper concentration in objects from Wadi Shab GAS-1, Ras Al-Hamra RH-10, Ras Al-Hadd HD-6, Ras Al-Jinz RJ-6, Ras Al-Hadd HD-10, Ras Al-Jinz RJ-2 and Ras Al-Jinz RJ-1, analyzed by XRF.

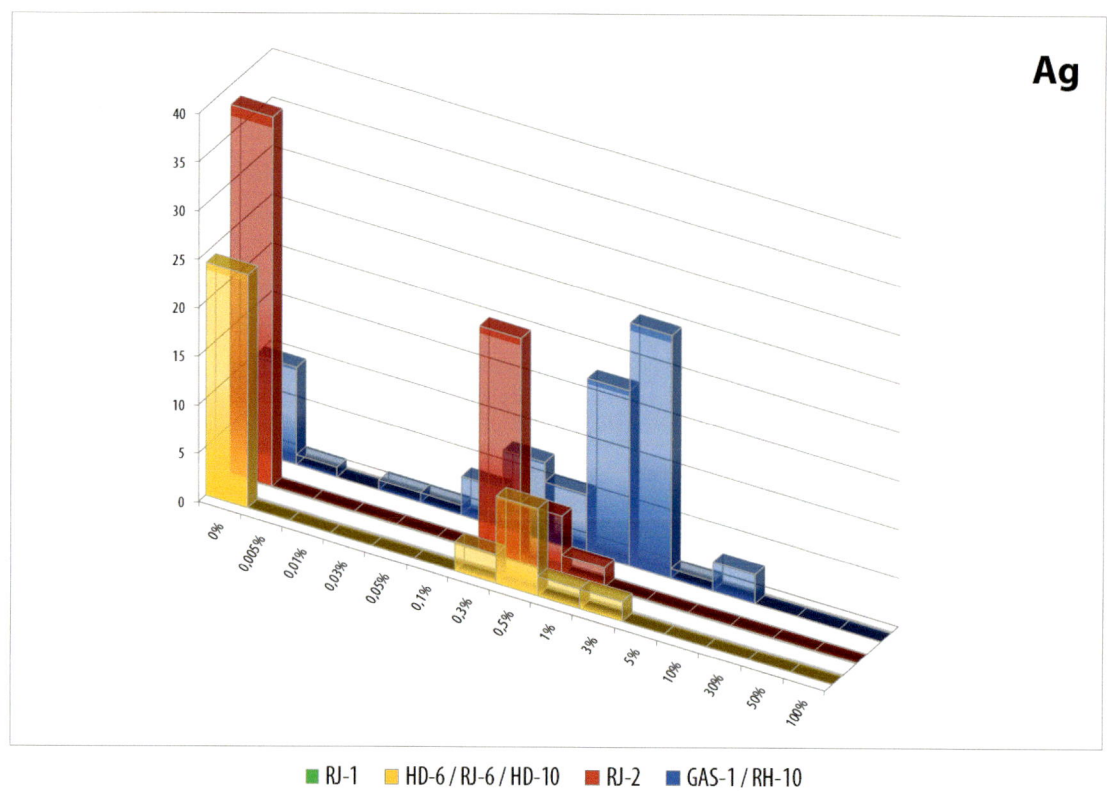

Figure 12.4. Silver concentration in objects from Wadi Shab GAS-1, Ras Al-Hamra RH-10, Ras Al-Hadd HD-6, Ras Al-Hadd HD-10, Ras Al-Jinz RJ-2 and Ras Al-Jinz RJ-6, analyzed by XRF.

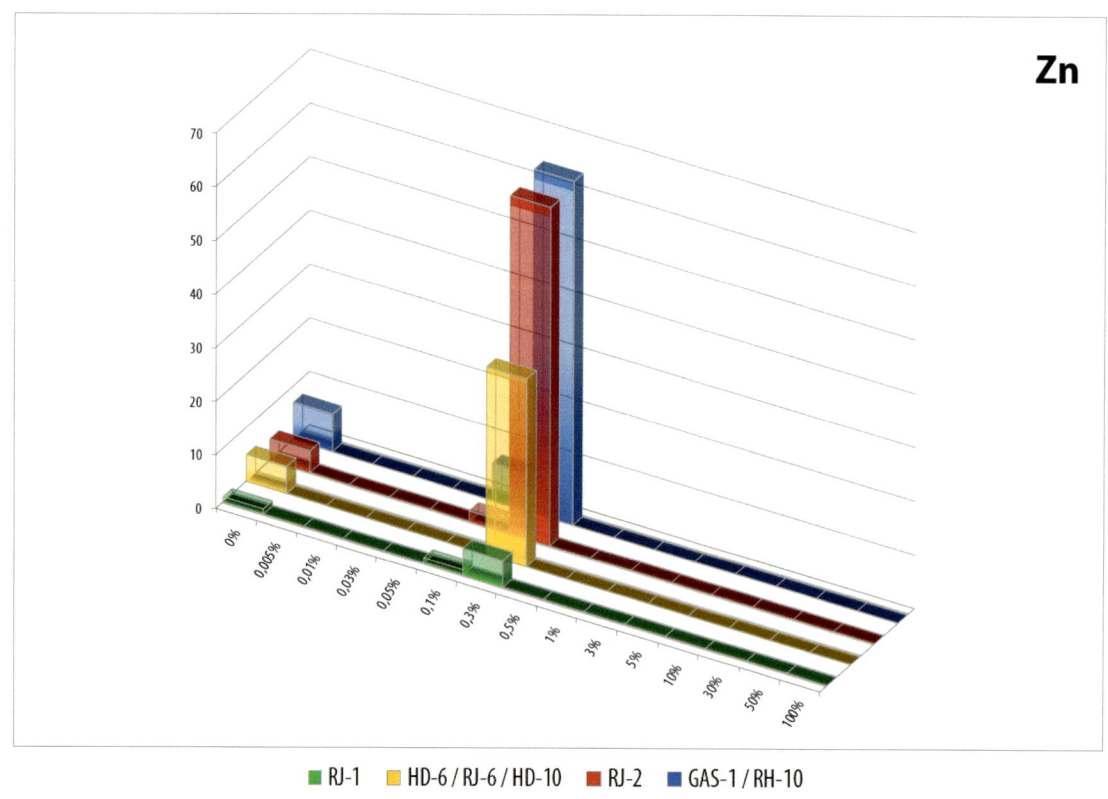

Figure 12.5. Zinc concentration in objects from Wadi Shab GAS-1, Ras Al-Hamra RH-10, Ras Al-Hadd HD-6, Ras Al-Hadd HD-10, Ras Al-Jinz RJ-6, Ras Al-Jinz RJ-2 and Ras Al-Jinz RJ-1, analyzed by XRF.

Chemical-Physical Analyses by Energy Dispersive X-Ray Fluorescence (EDXRF)

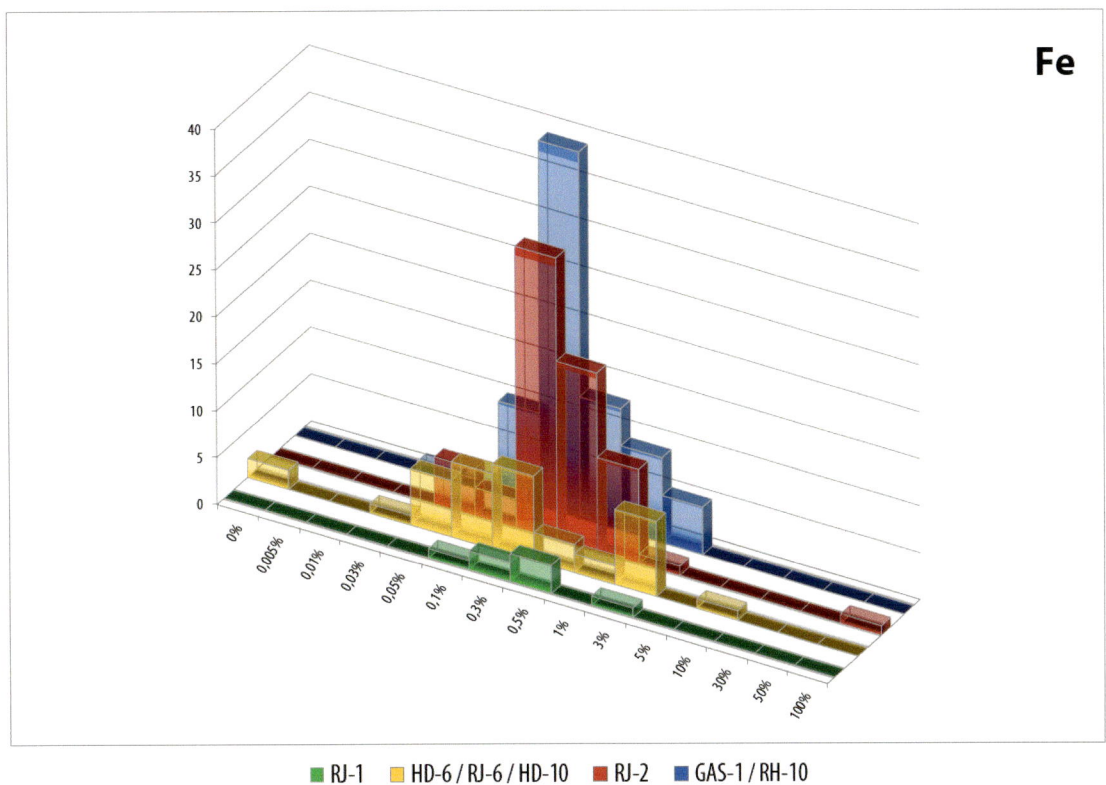

Figure 12.6. Iron concentration in objects from Wadi Shab GAS-1, Ras Al-Hamra RH-10, Ras Al-Hadd HD-6, Ras Al-Hadd HD-10, Ras Al-Jinz RJ-6, Ras Al-Jinz RJ-2 and Ras Al-Jinz RJ-1, analyzed by XRF.

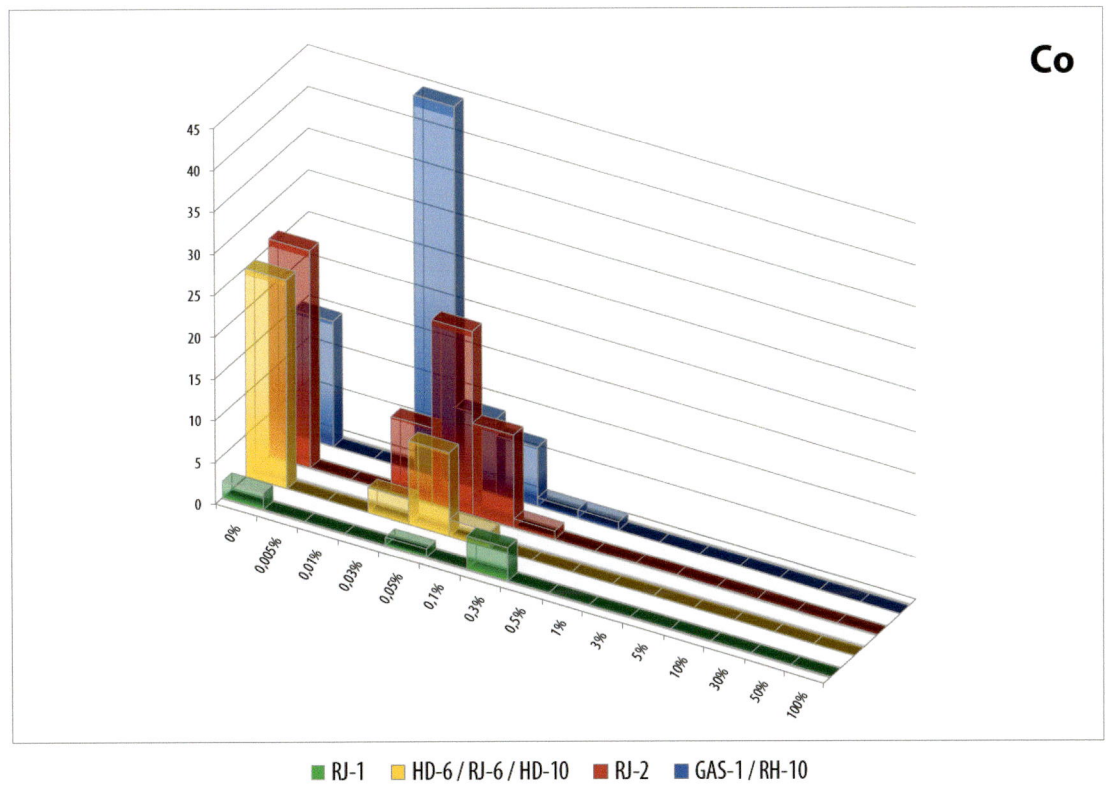

Figure 12.7. Cobalt concentration in objects from Wadi Shab GAS-1, Ras Al-Hamra RH-10, Ras Al-Hadd HD-6, Ras Al-Hadd HD-10, Ras Al-Jinz RJ-6, Ras Al-Jinz RJ-2 and Ras Al-Jinz RJ-1, analyzed by XRF.

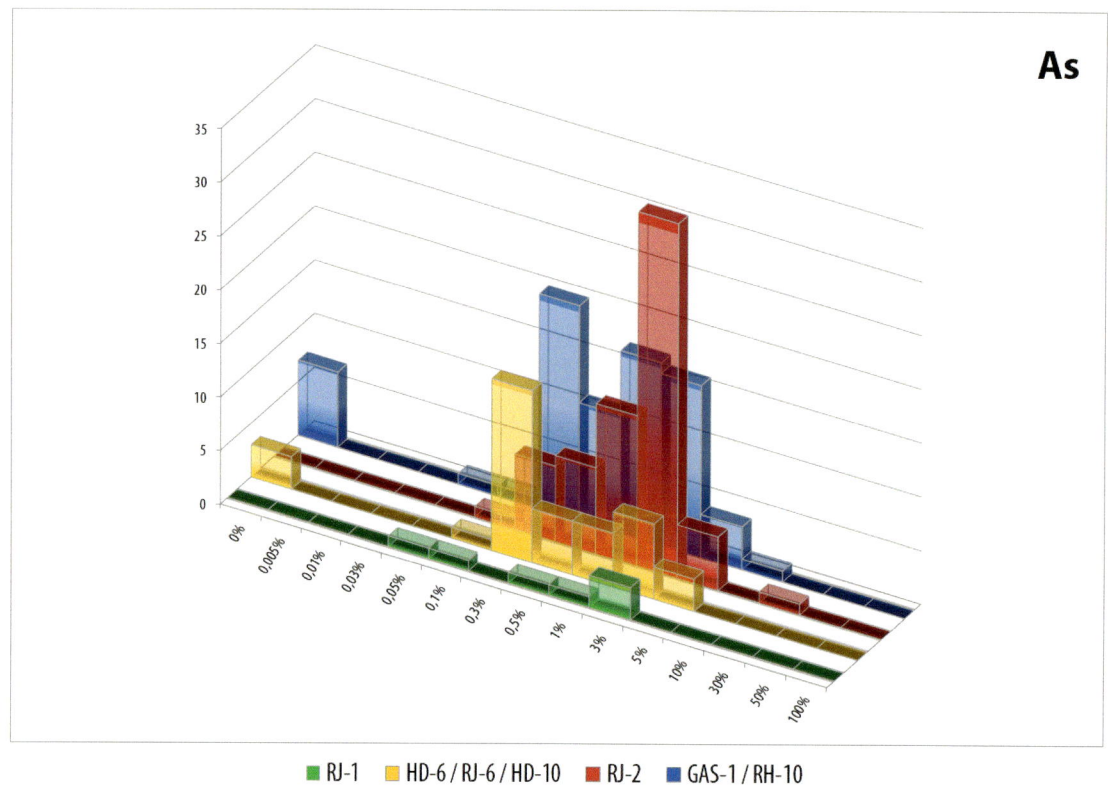

Figure 12.8. Arsenic concentration in objects from Wadi Shab GAS-1, Ras Al-Hamra RH-10, Ras Al-Hadd HD-6, Ras Al-Hadd HD-10, Ras Al-Jinz RJ-6, Ras Al-Jinz RJ-2 and Ras Al-Jinz RJ-1, analyzed by XRF.

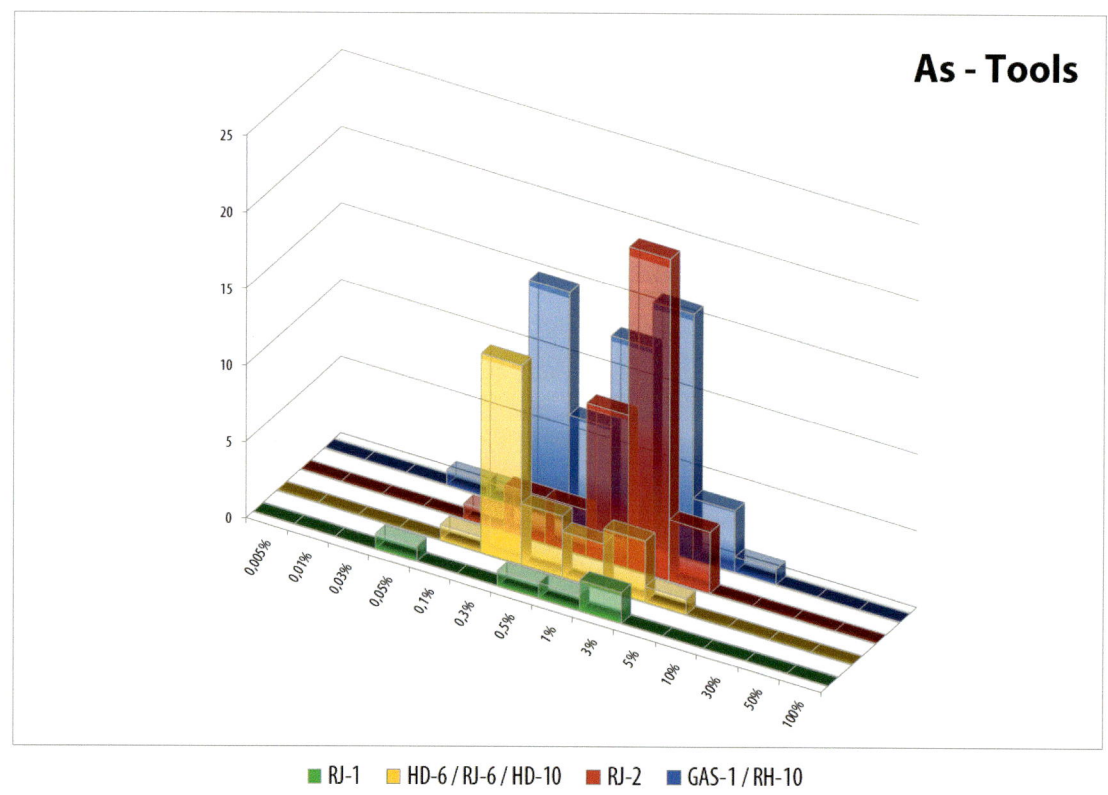

Figure 12.9. Arsenic concentration in tools from Wadi Shab GAS-1, Ras Al-Hamra RH-10, Ras Al-Hadd HD-6, Ras Al-Hadd HD-10, Ras Al-Jinz RJ-6, Ras Al-Jinz RJ-2 and Ras Al-Jinz RJ-1, analyzed by XRF.

Chemical-Physical Analyses by Energy Dispersive X-Ray Fluorescence (EDXRF)

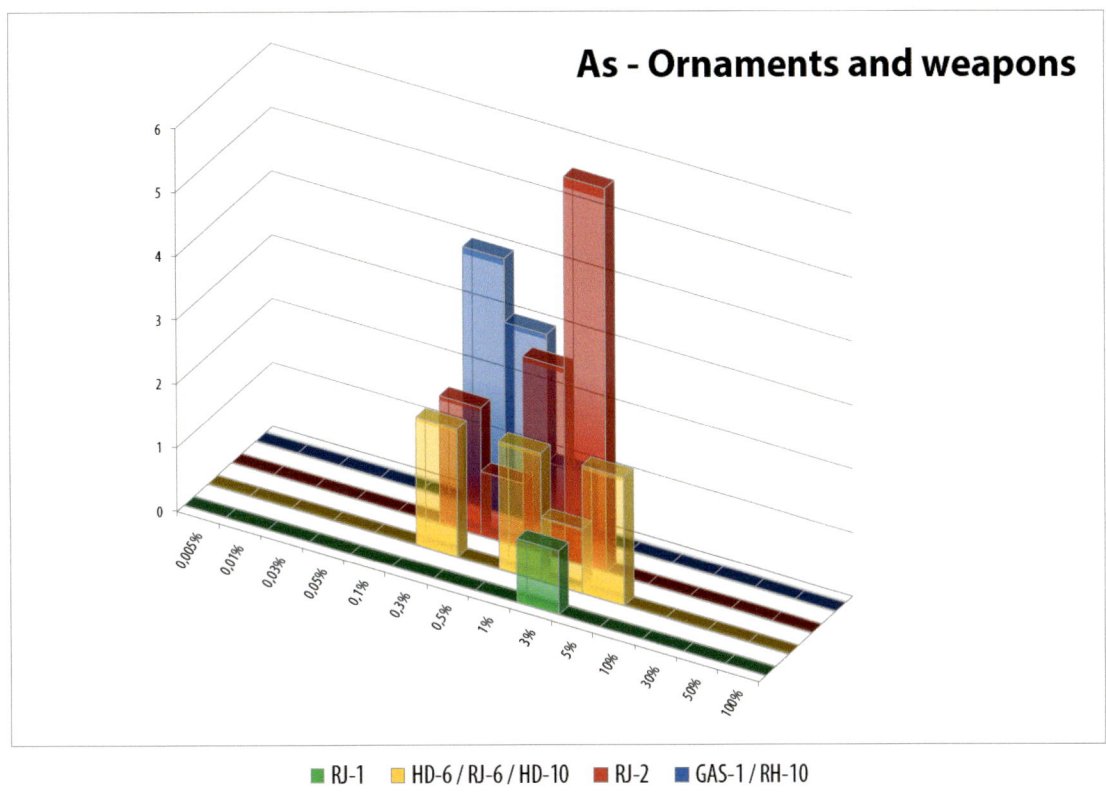

Figure 12.10. Arsenic concentration in ornaments and weapons from Wadi Shab GAS-1, Ras Al-Hamra RH-10, Ras Al-Hadd HD-6, Ras Al-Hadd HD-10, Ras Al-Jinz RJ-6 and Ras Al-Jinz RJ-2, analyzed by XRF.

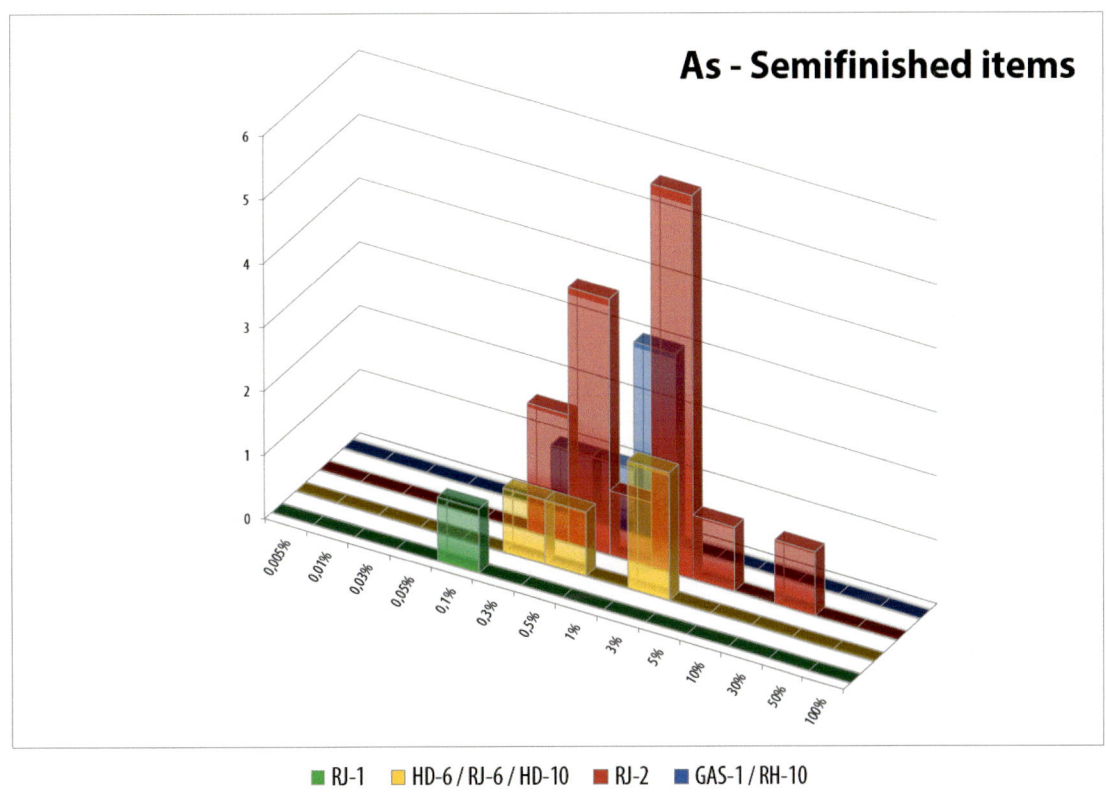

Figure 12.11. Arsenic concentration in semi-finished items from Wadi Shab GAS-1, Ras Al-Hamra RH-10, Ras Al-Hadd HD-6, Ras Al-Hadd HD-10, Ras Al-Jinz RJ-6, Ras Al-Jinz RJ-2 and Ras Al-Jinz RJ-1, analyzed by XRF.

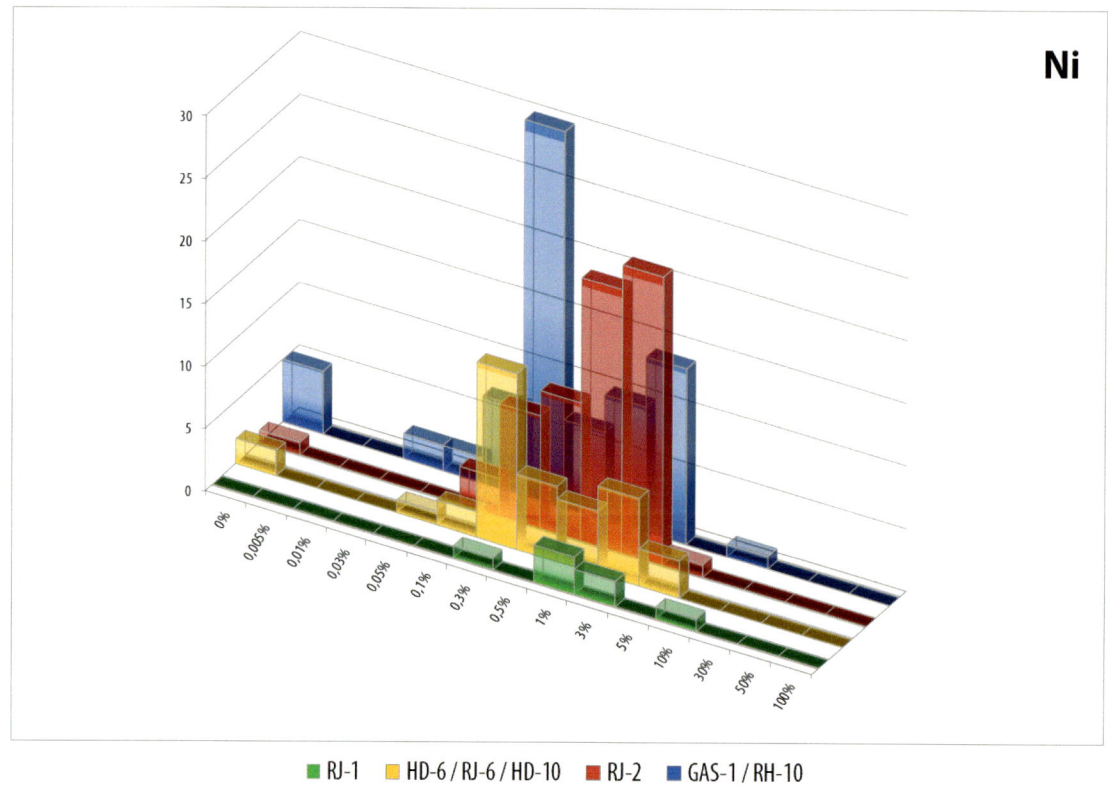

Figure 12.12. Nickel concentration in objects from Wadi Shab GAS-1, Ras Al-Hamra RH-10, Ras Al-Hadd HD-6, Ras Al-Hadd HD-10, Ras Al-Jinz RJ-6, Ras Al-Jinz RJ-2 and Ras Al-Jinz RJ-1, analyzed by XRF.

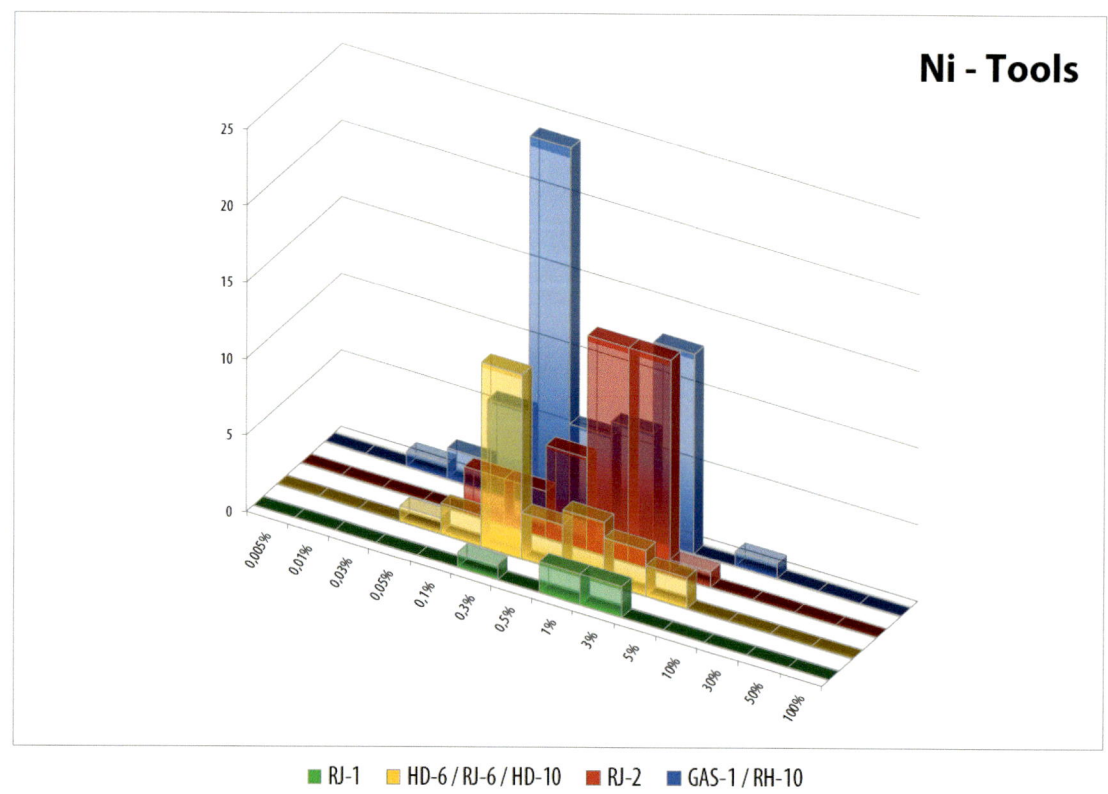

Figure 12.13. Nickel concentration in tools from Wadi Shab GAS-1, Ras Al-Hamra RH-10, Ras Al-Hadd HD-6, Ras Al-Hadd HD-10, Ras Al-Jinz RJ-6, Ras Al-Jinz RJ-2 and Ras Al-Jinz RJ-1, analyzed by XRF.

Chemical-Physical Analyses by Energy Dispersive X-Ray Fluorescence (EDXRF)

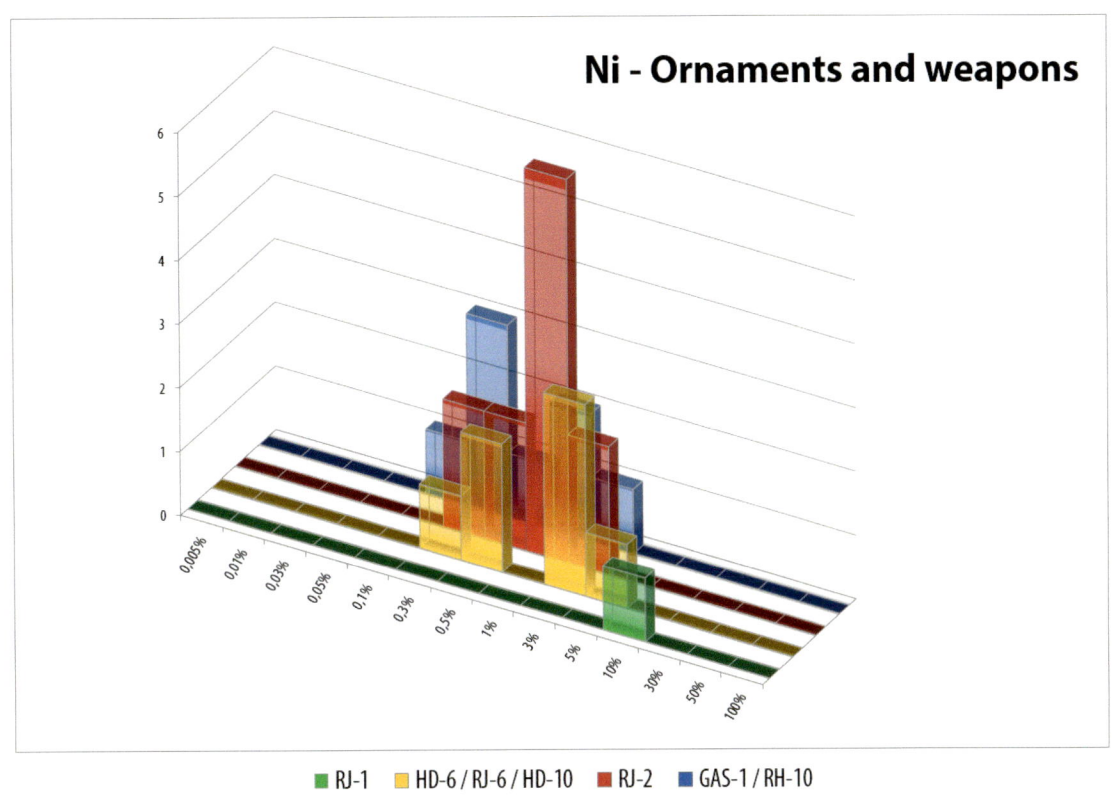

Figure 12.14. Nickel concentration in ornaments and weapons from Wadi Shab GAS-1, Ras Al-Hamra RH-10, Ras Al-Hadd HD-6, Ras Al-Hadd HD-10, Ras Al-Jinz RJ-6 and Ras Al-Jinz RJ-2 and Ras Al-Jinz RJ-1, analyzed by XRF.

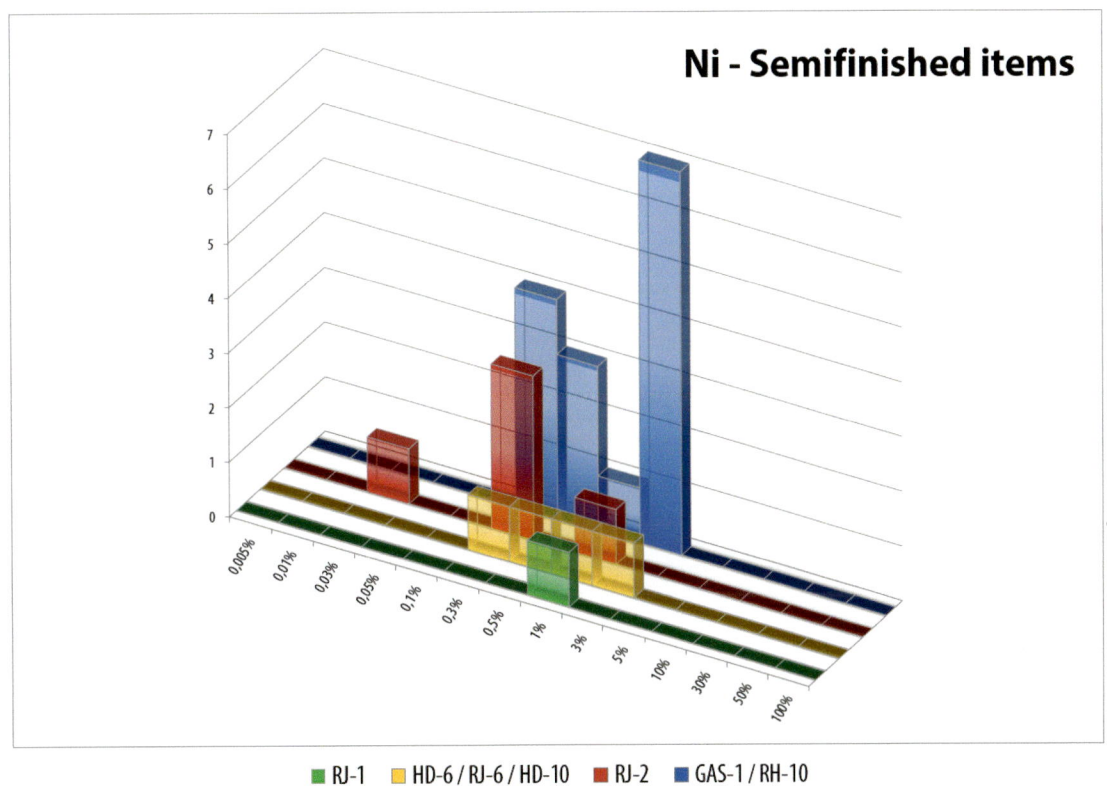

Figure 12.15. Nickel concentration in semi-finished items from Wadi Shab GAS-1, Ras Al-Hamra RH-10, Ras Al-Hadd HD-6, Ras Al-Hadd HD-10, Ras Al-Jinz RJ-6, Ras Al-Jinz RJ-2 and Ras Al-Jinz RJ-1, analyzed by XRF.

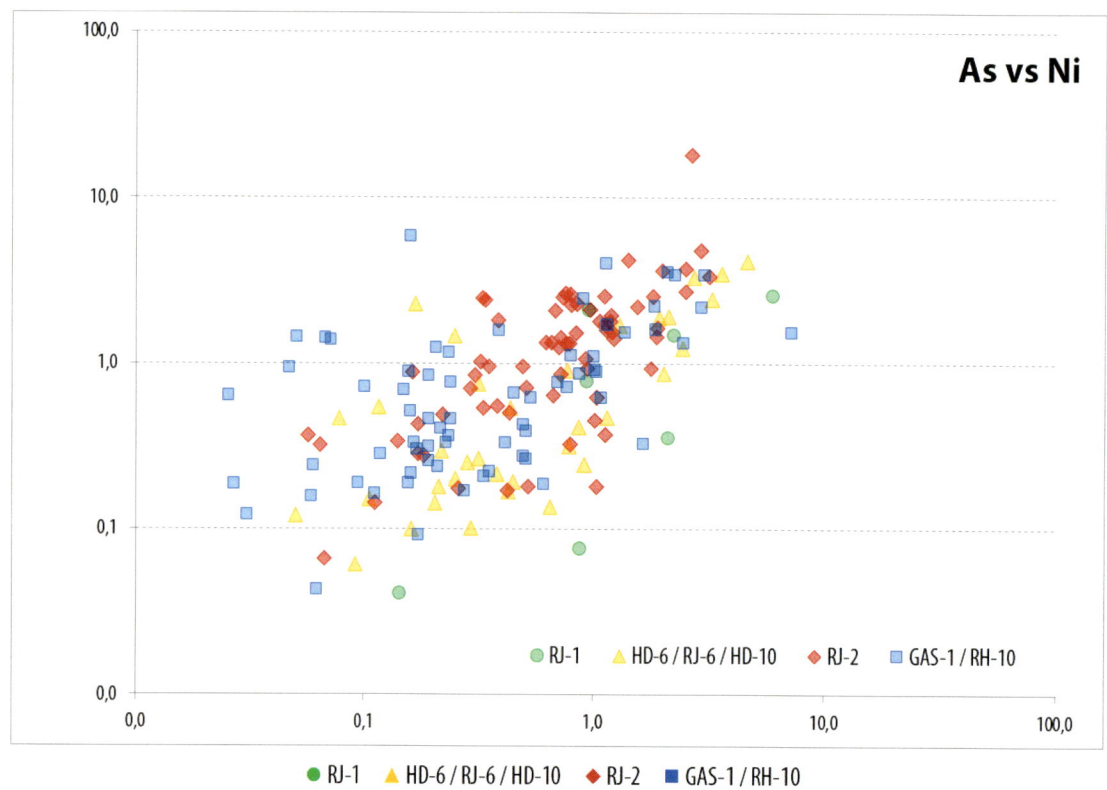

Figure 12.16. Arsenic and nickel in objects from Wadi Shab GAS-1, Ras Al-Hamra RH-10, Ras Al-Hadd HD-6, Ras Al-Hadd HD-10, Ras Al-Jinz RJ-6, Ras Al-Jinz RJ-2 and Ras Al-Jinz RJ-1, analyzed by XRF.

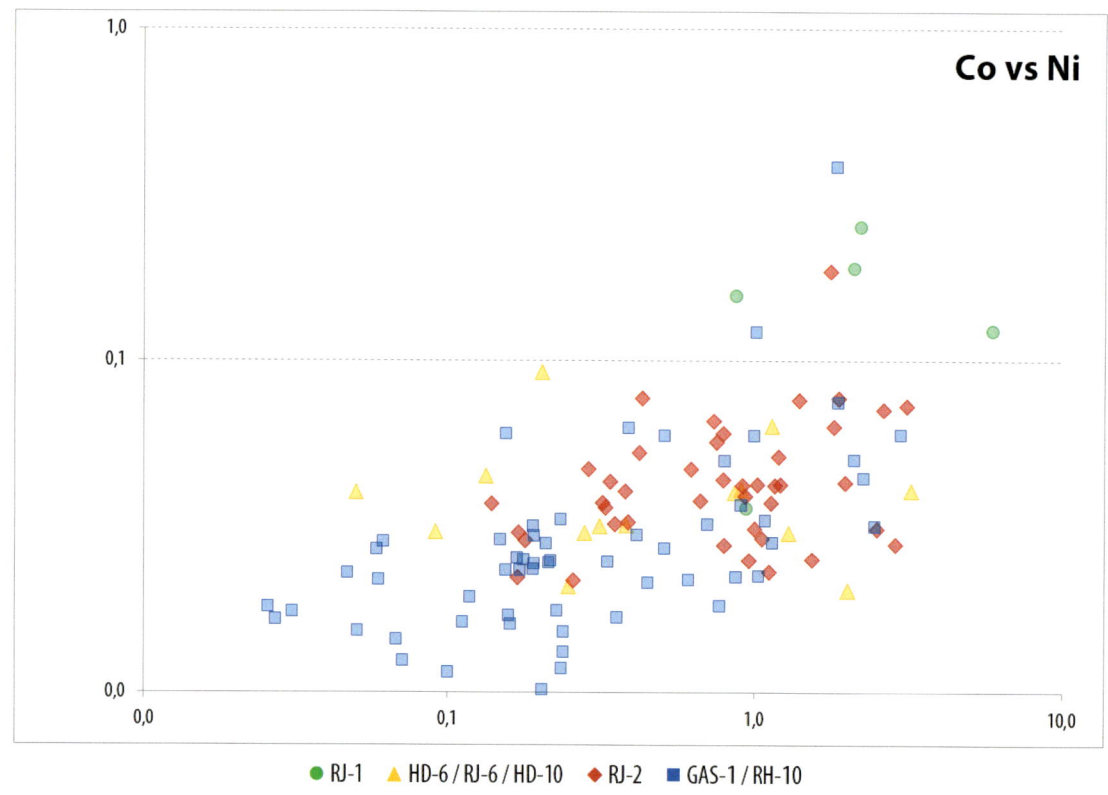

Figure 12.17. Cobalt and nickel in objects from Wadi Shab GAS-1, Ras Al-Hamra RH-10, Ras Al-Hadd HD-6, Ras Al-Hadd HD-10, Ras Al-Jinz RJ-6, Ras Al-Jinz RJ-2 and Ras Al-Jinz RJ-1, analyzed by XRF.

Chemical-Physical Analyses by Energy Dispersive X-Ray Fluorescence (EDXRF)

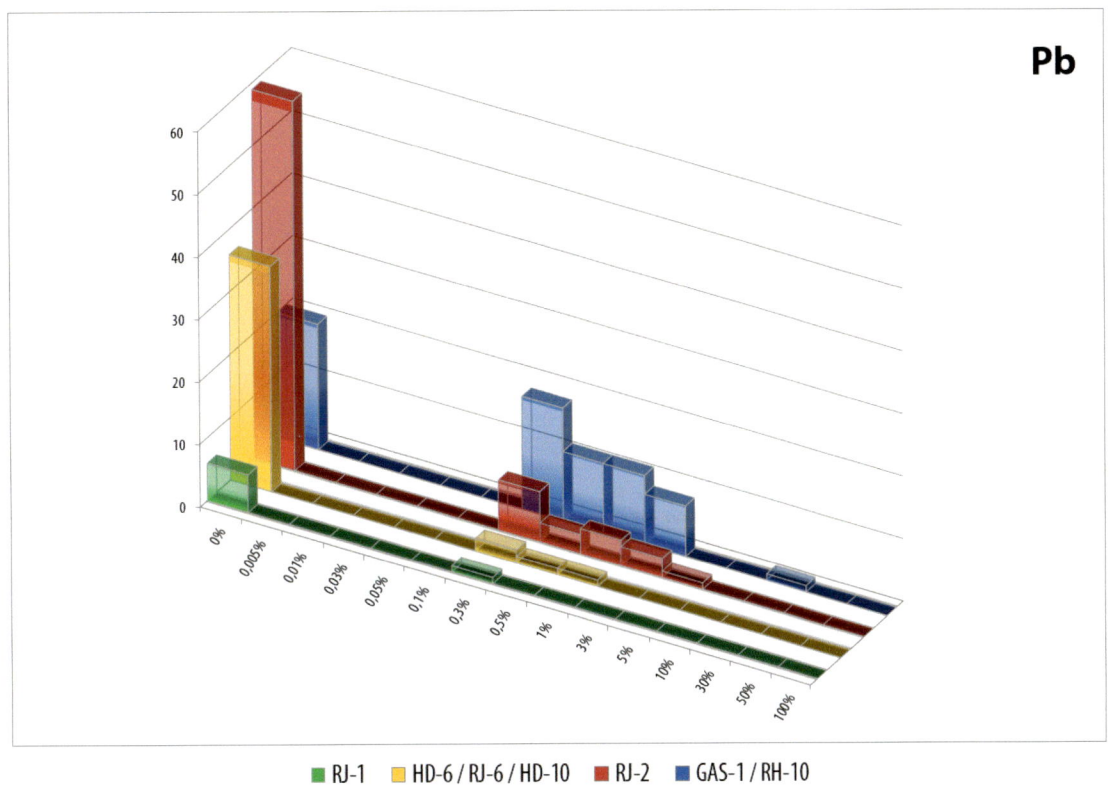

Figure 12.18. Lead concentration in objects from Wadi Shab GAS-1, Ras Al-Hamra RH-10, Ras Al-Hadd HD-6, Ras Al-Hadd HD-10, Ras Al-Jinz RJ-6, Ras Al-Jinz RJ-2 and Ras Al-Jinz RJ-1, analyzed by XRF.

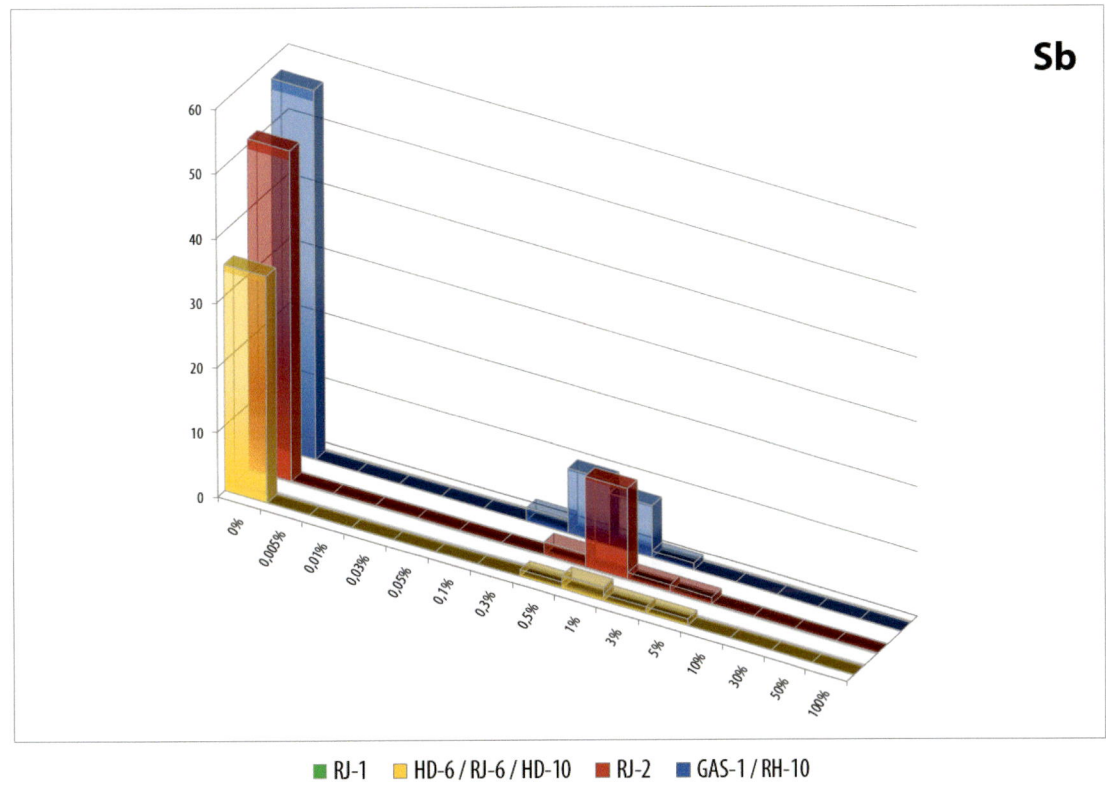

Figure 12.19. Antimony concentration in objects from Wadi Shab GAS-1, Ras Al-Hamra RH-10, Ras Al-Hadd HD-6, Ras Al-Hadd HD-10, Ras Al-Jinz RJ-2 and Ras Al-Jinz RJ-6, analyzed by XRF.

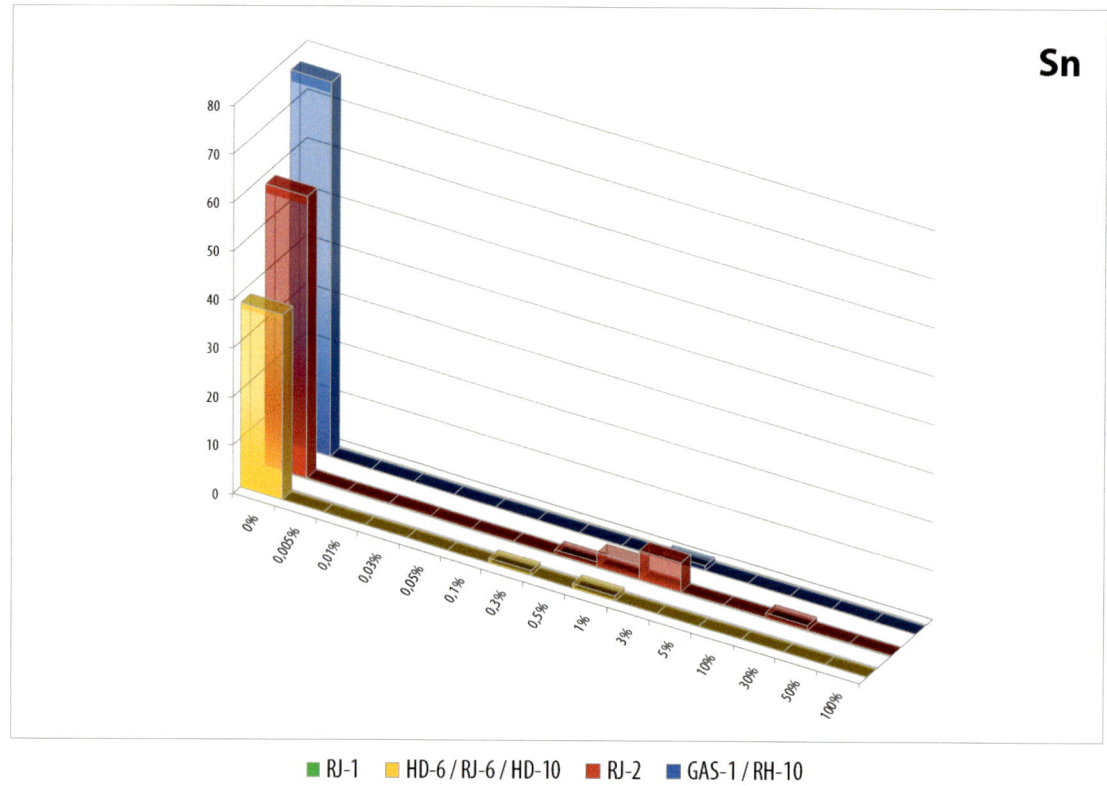

Figure 12.20. Tin concentration in objects from Wadi Shab GAS-1, Ras Al-Hamra RH-10, Ras Al-Hadd HD-6, Ras Al-Hadd HD-10, Ras Al-Jinz RJ-2 and Ras Al-Jinz RJ-6, analyzed by XRF.

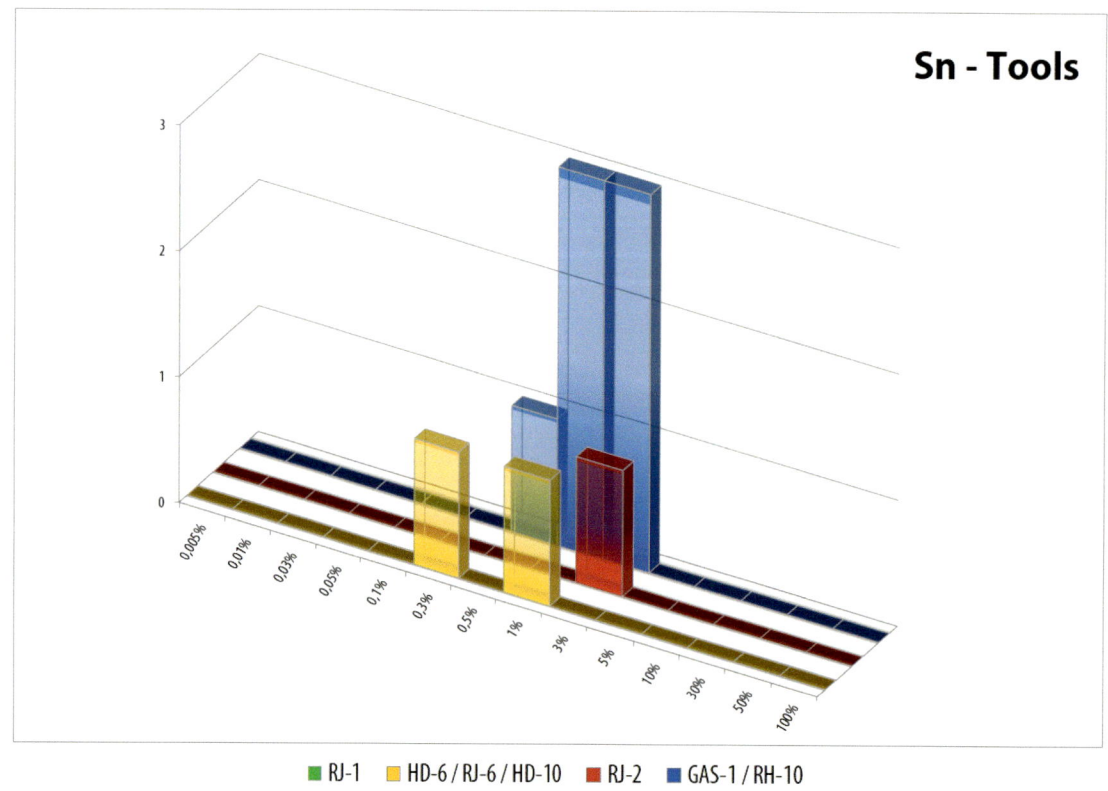

Figure 12.21. Tin concentration in tools from Wadi Shab GAS-1, Ras Al-Hamra RH-10, Ras Al-Hadd HD-6, Ras Al-Hadd HD-10, Ras Al-Jinz RJ-6, Ras Al-Jinz RJ-2, analyzed by XRF.

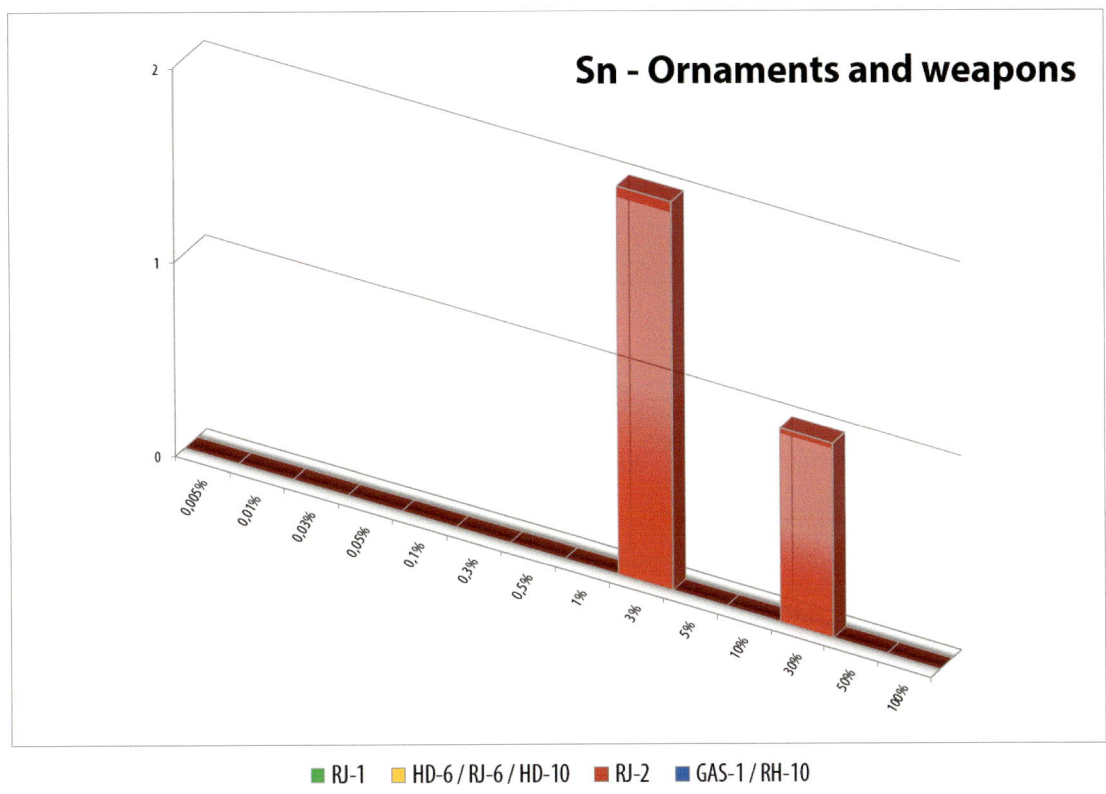

Figure 12.22. Tin concentration in ornaments and weapons from Ras Al-Jinz RJ-2, analyzed by XRF.

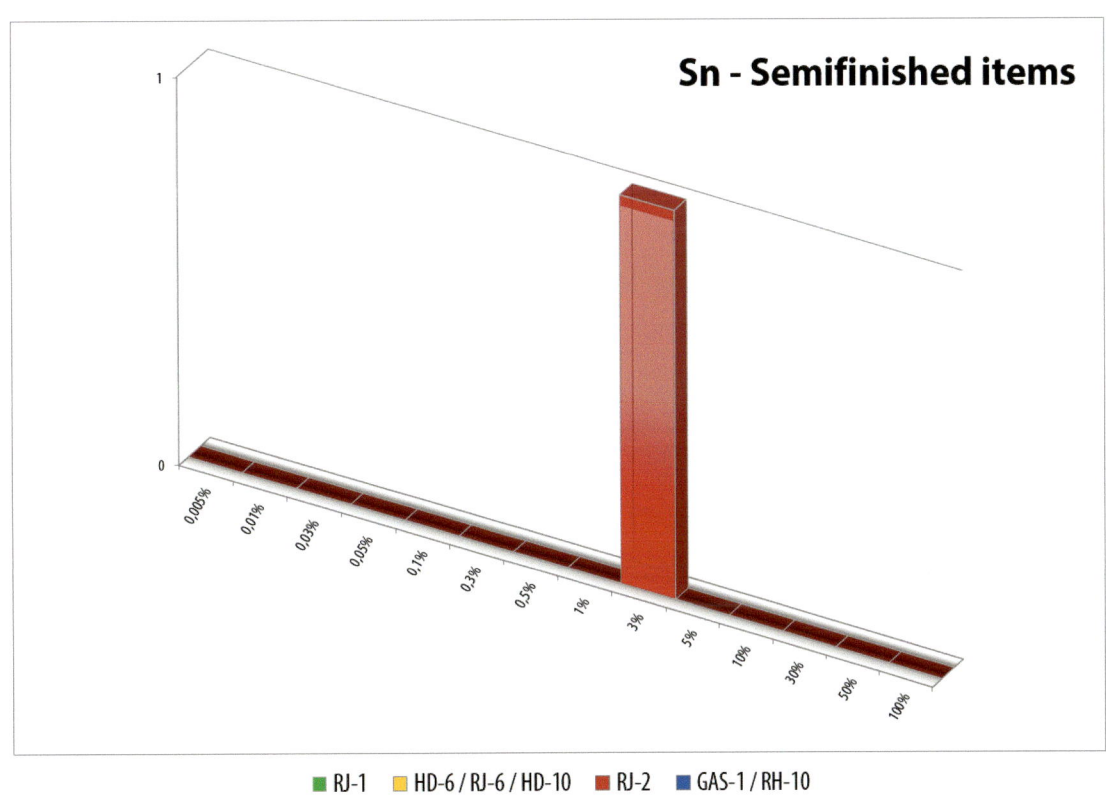

Figure 12.23. Tin concentration in semi-finished items from Ras Al-Jinz RJ-2, analyzed by XRF.

Bibliography

AA.VV., 2014. Penn Museum. Collections Database Search. Retrieved 19/09/2014 from: http://www.penn.museum/collections/object/119582.

ANDERSON, D. M. and W. L. Prell 1993. A 300 KYR Record of Upwelling Off Oman during the Late Quaternary, Evidence of the Asian Southwest Monsoon. *Paleoceanography* 8.2, 193–208.

ANDRÉ-SALVINI, B. 2002. "The land where the sun rises…" The Representation of Dilmun in Sumerian Literature. In Crawford H. E. W. and M. Rice (eds.), *Traces of Paradise. The Archaeology of Bahrain 2500 BC – 300 AD*. London, I. B. Tauris & Company, pp. 28–34.

AL-SHANFARI, A. B. and G. Weisgerber 1989. A Late Bronze Age Warrior Burial from Nizwa (Oman). In Costa P. M. and M. Tosi (eds.), *Oman Studies. Papers on the Archaeology and History of Oman* (Serie Orientale Roma LXIII). Rome, Istituto Italiano per il Medio ed Estremo Oriente, pp. 17–30.

BALDACCI, M. 2011. *La marineria nella Mesopotamia antica* (Rivista Marittima, Supplemento). Rome, Ministero della Difesa della Repubblica Italiana.

BEGEMANN, F., A. Hauptmann, S. Schmitt-Strecker and G. Weisgerber 2010. Lead isotope and chemical signature of copper from Oman and its occurrence in Mesopotamia and sites on the Arabian Gulf coast. *Arabian Archaeology and Epigraphy* 21, 135–169.

BEGEMANN, F. and S. Schmitt-Strecker 2009. Über das früe Kupfer Mesopotamiens. *Iranica Antiqua* 44, 1–45.

BERTHOUD, T. 1979. *Etude par l'analyse de traces et la modélisation de la filiation entre minerai de cuivre et objects archéologiques du Moyen-Orient (IVème et IIIème millénaire avant notre ère)* (Unpublished doctoral dissertation). University Paris VI "Pierre et Marie Curie".

BIBBY, T. G. 1970. *Looking for Dilmun*. New York, Knopf.

BIBBY, T. G. 1996. *Looking for Dilmun* (New Edition). London, Stacey International.

BISWAS, A. K. 1996. *Minerals and Metals in Ancient India, Volume 1. Archaeological Evidence*. New Delhi, D. K. Printworld.

BORTOLINI, E. and M. Tosi 2011. Dal Kinship al Kinship, Le tombe collettive nell'Oman del terzo millennio a.C. e la costruzione della civiltà di Magan. In Nizzo V. (Editor), *Antropologia e archeologia a confronto, Dalla nascita alla morte, antropologia e archeologia a confronto. Atti dell'Incontro Internazionale di studi in onore di Claude Lévi-Strauss*. Rome, Editorial Service System, pp. 287–317.

CABLE, C. M. and C. P. Thornton 2013. Monumentality and the third-millennium 'towers' of the Oman Peninsula. In Abraham S., P. Gullapalli, T. Raczek and U. Rizvi (eds.), *Connections and Complexity, New Approaches to the Archaeology of South Asia*. Walnut Creek, CA, Left Coast Press, pp. 375–399.

CHAKRABARTI, D. K. and N. Lahiri 1996. *Copper and its Alloys in Ancient India*. New Delhi, Munishiram Manoharlal Publishers Pvt Ltd.

CHENG, C. F. and C. M. Schwitter 1957. Nickel in ancient bronzes. *American Journal of Archaeology* 61, 351–365.

CLEUZIOU, S. 2000. Dilmun, Origins and Early Development. In Crawford H. E. W. and M. Rice (eds.), *Traces of Paradise. The Archaeology of Bahrain 2500 BC – 300 AD*. London, I. B. Tauris & Company, pp. 24–27.

CLEUZIOU, S. and M. Tosi 1994. Black boats of Magan. Some thoughts on Bronze Age water transport in Oman and beyond from the impressed bitumen slabs of Ra's al-Junayz. In Parpola A. and P. Koskikallio (eds.), *South Asian Archaeology 1993, Volume 2* (Annales Academiae Scientiarum Fennicae series B, 271). Helsinki, Suomalainen Tiedeakatemia, pp. 745–761.

CLEUZIOU, S. and M. Tosi 2000. Ra's al-Jinz and the Prehistoric Coastal Cultures of the Ja'alān. *Journal of Oman Studies* 11, 19–73.

CLEUZIOU, S. and M. Tosi 2007. *In the Shadow of the Ancestors. The Prehistoric Foundations of the Early Arabian Civilization in Oman*. Muscat, Ministry of Heritage and Culture of the Sultanate of Oman.

COLEMAN, R. G. and E. H. Bailey 1981. *Oman Mineral Deposits and Geology of Northern Oman as of 1974* (U.S. Geological Survey, Open-File Report 81-452). Muscat, Directorate General of Petroleum and Minerals, Ministry of Development of the Sultanate of Oman.

COLTORTI, M. 1989. Geomorphological characteristics of the Ras al-Junaiz area (Sultanate of Oman). In Costa P. M. and M. Tosi (eds.), *Oman Studies. Papers on the Archaeology and History of Oman* (Serie Orientale Roma LXIII). Rome, Istituto Italiano per il Medio ed Estremo Oriente, pp. 79–96.

CONSTANTINOU, G. 1982. Geological Features and Ancient Exploitation of the Cupriferous Sulphide Orebodies of Cyprus. In Muhly J.D., R. Maddin and V. Karageorghis (eds.), *Early Metallurgy in Cyprus 4000–500 BC. Acta of the International Archaeological Symposium (Larnaca, Cyprus 1981)*. Nicosia, Pierides Foundation, Department of Antiquities Republic of Cyprus, pp. 13–24.

CRADDOCK, P.T. 1985. Technical Appendix 1. The composition of metal artefacts. *Oriens Antiquus* 24, 97–101.

CRADDOCK, P.T. 1995. *Early Metal Mining and Production*. Edinburgh, Edinburgh University Press.

CRADDOCK, P. T., S. La Niece and D. R. Hook 2003. Evidence for the Production, Trading and refining of Copper in the Gulf of Oman during the Third millennium BC. In Stöllner T., Körlin G., Steffens G. and J. Cierny (eds.), *Man and Mining – Mensch und Bergbau. Studies in Honour of Gerd Weisgerber in occasion of his 65th birthday* (Der Anschnitt 16). Bochum, Bergbau-Museum, pp, 103–112.

CRAWFORD, H. 2000. Bahrain, Warehouse of the Gulf. In Crawford H. E. W. and M. Rice (eds.), *Traces of Paradise. The Archaeology of Bahrain 2500 BC – 300 AD*. London, I. B. Tauris & Company, 72–86.

DE JESUS, P.S. 1980. *The development of prehistoric mining and metallurgy in Anatolia* (BAR International Series 74). Oxford, Archaeopress.

DEL MONTE, G. 2004. *Iscrizioni reali dal Vicino Oriente antico*. Pisa, Università di Pisa

ECKSTEIN, D., W. Liese and J. Stieber 1987. *Holzversorgungen im prähistorischen Kupferbergbau in Oman*. Naturwissenschaftliche Rundschau 40.11, 426–430.

FRENKEN, K. (Editor) 2009. *Irrigation in the Middle East region in figures* (AQUASTAT Survey 2008). Rome, Food and Agriculture Organization of the United Nations.

FRIFELT, K. 1991. T*hird Millennium Graves. The Island of Umm An-Nar 1* (Jutland Archaeological Society Publications XXVI, 1). Aarhus, Aarhus University Press.

FRIFELT, K. 1995. *The Third Millennium Settlement. The Island of Umm an-Nar 2* (Jutland Archaeological Society Publications XXVI, 2). Aarhus, Jutland Archaeological Society.

GALE, N. H. 1991. Copper Oxhide Ingots. Their origin and their place in the Bronze Age metals trade in the Mediterranean. In Gale N.H. (eds.), *Bronze Age Trade in the Mediterranean* (Studies in Mediterranean Archaeology 90). Jonsered, Paul Åströms Förlag, pp. 197–239.

GALE, N. H. and Z. A. Stos-Gale 2000. Lead isotope analyses applied to provenance studies. In Ciliberto E. and G. Spoto (eds.), *Modern Analytical Methods in Art and Archaeology*. New York, Wiley, pp. 503–584.

GIARDINO, C. 1995. *Il Mediterraneo occidentale fra XIV ed VIII secolo a.C. Cerchie minerarie e metallurgiche. The West Mediterranean between the 14th and 8th Centuries B.C. Mining and metallurgical spheres* (BAR International Series 612). Oxford, Tempus Reparatum.

GIARDINO, C. 2000. Prehistoric copper activity at Pyrgos. *Report of the Department of Antiquities of Cyprus*, 19–32.

GIARDINO, C. 2008. Strumenti e tecniche tradizionali per la produzione del vasellame metallico. Il ramaio di Roccagorga (Latina) e l'archeometallurgia. In Lugli F. and A. A. Stoppiello (eds.), *Atti del 3° Convegno Nazionale di Etnoarcheologia (Mondaino 2004), Proceedings of the 3rd Italian Congress of Ethnoarchaeology (Mondaino, 2004)* (BAR International Series 1841). Oxford, Archaeopress, pp. 27–35.

GIARDINO, C. 2010. *I metalli nel mondo antico. Introduzione all'archeometallurgia*. Roma-Bari, Laterza.

GIARDINO, C. 2011. Indagini archeometallurgiche sui reperti. In Gran-Aymerich J. and A. Domínguez-Arranz (eds.), *La Castellina a sud di Civitavecchia. Origini ed eredità. Origines protohistoriques et évolution d'un habitat étrusque*. Roma, L'Erma di Bretschneider, 981–998.

GIARDINO, C., G. Guida and S. Ridolfi 2007. The prehistoric metals from Oman: new data. *Archaeometallurgy in Europe (2nd International Conference 2007)*. Milan, AIM, 1–8.

GIARDINO, C. and A. Lazzari 2014. Bronze Age Metal Manufacturing in Eastern Arabia. Evidence from Ra's al-Jinz (Oman) and Failaka (Kuwait). In Lamberg–Karlovsky C.C., B. Genito and B. Cerasetti (eds.), *'My life is like the summer rose'. Maurizio Tosi e l'archeologia come modo di vivere: papers in honour of Maurizio Tosi for his 70th birthday* (BAR international series, 2690). Oxford, Archaeopress, pp. 311–322.

GIARDINO, C. and G. Paternoster, in press. Chemical-Physical Analyses by Energy Dispersive X-Ray Fluorescence (ED-XRF) of metal finds. In Yule P. A. and G. Gernez (eds.), *Early Iron Age Metal-Working Workshop in the Empty Quarter, Sultanate of Oman*. Bonn, Rudolf Habelt GmbH.

GLASSNER, J. J. 1989. Mesopotamian textual evidence on Magan/Makan in the late 3rd Millennium B.C. In Costa P. M. and M. Tosi (eds.), *Oman Studies. Papers on the Archaeology and History of Oman* (Serie Orientale Roma LXIII). Rome, Istituto Italiano per il Medio ed Estremo Oriente, pp. 181–191.

GOSTENČNIK, K. 2010. The Magdalensberg Textile Tools. A Preliminary Assessment. In Andersson Strand E., M. Gleba, U. Mannering, C. Munkholt and M. Ringgaard (eds.), *North European Symposium for Archaeological Textiles X* (Ancient Textiles Series 5). Oxford, Oxbow Book, pp. 73–90.

HASTINGS, A., J. H. Humphries and R. H. Meadow 1975. Oman in the third millennium BCE. *Journal of Oman Studies* 1, 9–55.

HAUPTMANN, A. 1985. *5000 Jahre Kupfer in Oman / 1. Die Entwicklung der Kupfermetallurgie vom 3. Jahrtausend bis zur Neuzeit* (Der Anschnitt 4). Bochum, Deutsches Bergbau Museum.

HAUPTMANN, A. 1995. Chemische Zusammensetzung von Metallobjecten aus der Siedlung Umm an-Nar. In Frieflet K. (eds.), *The Island of Umm an-Nar Vol. 2. The Third Millennium Settlement* (Jutland Archaeological Society Publications 26/2). Aarhus, Jutland Archaeological Society, pp. 246–248.

HAUPTMANN, A. and G. Weisgerber 1981. Third millennium BC copper production in Oman. *Revue d'Archeometrie* 1.1, 131–138.

HAUPTMANN, A., G. Weisgerber and H. G. Bachmann 1988. Early copper metallurgy in Oman. In Maddin R. (eds.), *The Beginning and the Use of Metals and Alloys. Papers from the Second International Conference on the Beginning and the Use of Metals and Alloys (Zhenzhou, China 1986)*. Cambridge MA, MIT Press, pp. 34–51.

HEIMPEL, W. 1982. A first step in the diorite question. *Revue d'Assyriologie et d'Archéologie Orientale* 76.1, 65–67.

HEIMPEL, W. 1987. Das Untere Meer. *Zeitschrift für Assyriologie* 77, 23–91.

HESKEL, D. L. 1983. A Model for the Adoption of Metallurgy in the Ancient Middle East. *Current Anthropology* 24.3, 362–366.

HØJLUND, F. 2000. Qal'at al Bahrain in the Bronze Age. In Crawford H. E. W. and M. Rice (eds.), *Traces of Paradise. The Archaeology of Bahrain 2500 BC – 300 AD*. London, I. B. Tauris & Company, pp. 59–62.

HØJLUND, F. and H. Andersen (eds.) 1994. *Qala'at al Bahrain / 1. The Northern City Wall and the Islamic Fortress* (Jutland Archaeological Society Publications 30.1). Hojbjerg, Jutland archaeological society.

HOLZER, H.F., M. Momenzadeh and G. Gropp 1971. Ancient Copper Mines in the Veshnoveh Area, Kuhestan-E-Qom, West-Central Iran. *Archaeologia Austriaca* 49, 1–22.

JASTROW, M. 1915. *The Civilization of Babylonia and Assyria: its remains, language, history, religion, commerce, law, art, and literature.* Philadelphia / London, J. B. Lippincott.

JONES, A. and G. MacGregor 2002. Introduction. Wonderful things: colour studies in archaeology from Munsell to materiality. In Jones A. and G. MacGregor (eds.), *Colouring the Past*. Oxford, New York, Berg, pp. 1–21.

KENNY, N. 2010. *Experimental Charcoal Making*. Retrieved on 02/09/2014 from: http://charcoal.seandalaiocht.com

KENNY, N. and B. Dolan n.d. *Traditional Charcoal Making. Experimental Archaeology*. Retrieved 02/09/2014 from: http://charcoal.seandalaiocht.com/uploads/3/5/0/8/3508898/leaflet_on_trad_charcoal_making_for_experiment.pdf

KENOYER, J. M. and H. M.-L. Miller 1999. Metal Technologies of the Indus Valley Tradition in Pakistan and Western India. In V. C. Pigott (ed.), *The Archaeometallurgy of the Asian Old World*. Philadelphia, University of Pennsylvania Museum, pp. 107–151.

KILLICK, R. and J. Moon 2005. *The Early Dilmun Settlement at Saar*. Ludlow, Archaeology International.

KINNAIRD, J. A. and P. Bowden 1991. Magmatism and Mineralization Associated with Phanerozoic Anorogenic Plutonic Complexes of the African Plate. In Kampunzu A. B. and R. T. Lubala (eds.), *Magmatism in extensional structural settings: the Phanerozoic African Plate*. Berlin / New York, Springer-Verla, pp. 410–487.

KNAPP, A. B. 2013. T*he Archaeology of Cyprus. From Earliest Prehistory through the Bronze Age*. Cambridge, Cambridge University Press.

KRAMER, N. S. 1970. *The Sumerians*. Chicago, University of Chicago Press.

KROLL, S. 1981. Der prähistorische Schlackenplatz Bir Kalher. In Weisgerber G. (ed), Mehr als Kupfer in Oman. Ergebnisse der Expedition 1981, Der Anschnitt. Zeitschrift für Kunst und Kultur im Bergbau 33.5-8, pp. 210–211.

LE MÉTOUR, J., J. C. Michel, F. Béchennec, J. P. Platel and J. Roger 1995. *Geology and Mineral Wealth of the Sultanate of Oman*. Muscat, Ministry of Petroleum and Minerals, Directorate General of Minerals.

LEIGH, B. 2016. Metal (Chapter 11). In Thornton C. P., C. M. Cable and G. L. Possehl (eds.), *The Bronze Age Towers at Bat, Sultanate of Oman. Research by the Bat Archaeological Project, 2007–12*. Philadelphia, University of Pennsylvania Press, pp. 229-238.

LLEWELLYN-SMITH, R. 2014. *Deserts and xeric shrublands. Southwestern Asia, most of Saudi Arabia, extending into Oman, United Arab Emirates, Yemen, Egypt, Iraq, Jordan, and Syria*. Retrieved 02/09/2014 from: http://www.worldwildlife.org/ecoregions/pa1303.

LO SCHIAVO, F. 2009. The Oxhide Ingots in Nuragic Sardinia. In Lo Schiavo F., J. D. Muhly, R. Maddin and A. Giumlia-Mair (eds.), *Oxhide Ingots in the Central Mediterranean*. Rome, A.G. Leventis Foundation / CNR-Istituto di studi sulle Civiltà dell'Egeo e del Vicino Oriente, pp. 225–390.

LOMBARD, P. 2000. Early Dilmun Burial Offerings. In Crawford H. E. W. and M. Rice (eds.), *Traces of Paradise. The Archaeology of Bahrain 2500 BC – 300 AD*. London, I. B. Tauris & Company, pp. 42–58.

LOMBARD, P. and M. Kervran 1989. *Bahrain National Museum Archaeological Collections: A selection of pre-Islamic antiquities*. Bahrain, Directorate of Museums and Heritage, Ministry of Information.

MANDAVILLE, J.P. 1986. Plant life in the Rub' al-Khali (the Empty Quarter), south-central Arabia. *Proceedings of the Royal Society of Edinburgh* (Section B: Biological Sciences), 147–157.

MARYON, H. 1971. *Metalwork & Enameling* (5th Revised Edition). New York, Dover Publications.

MERRILLEES, R. S. 1984. Ambelikou-Aletri. A Preliminary Report. *Report of the Department of Antiquities of Cyprus*, 1–13.

MICHEL, J. C. 1993 (Editor). *Mineral Occurrence and Metallogenic Map of North Oman (Scale 1, 500,000)*. Muscat, Ministry of Petroleum and Minerals, Directorate General of Minerals.

MILLER, N. F. 1984. The use of dung as fuel. An ethnographic example and an archaeological application. *Paléorient* 10.2, 71–79.

MORTIMER, A. 2016. Tower 1156 (Chapter 6). In Thornton C. P., C. M. Cable and G. L. Possehl (eds.), *The Bronze Age Towers at Bat, Sultanate of Oman. Research by the Bat Archaeological Project, 2007–12*. Philadelphia, University of Pennsylvania Press, pp. 123-153.

MUHLY, J. D. 2005. Cyprus and Copper for the World. In Yalçin Ü. (ed.), *Anatolian Metal III* (Der Anschnitt 18). Bochum, Bergbau-Museum, pp. 137–141.

MUHLY, J. D. 2009. Oxhide Ingots in the Aegean and in Egypt. In Lo Schiavo F., J. D. Muhly, R. Maddin and A. Giumlia-Mair (eds.), *Oxhide Ingots in the Central Mediterranean*. Rome, A.G. Leventis Foundation / CNR-Istituto di studi sulle Civiltà dell'Egeo e del Vicino Oriente, pp. 17–33.

NORTON, J., S. Abdul Majid, D. Allan, M. Al-Safran, B. Böer and R. Richer 2009. *An Illustrated Checklist of the Flora of Qatar*. Gosport, UK, Browndown Publications.

ORSI, P. 1906. Gela. Scavi del 1900–1905. *Monumenti Antichi dell'Accademia dei Lincei* 17, 5–766.

OTTAWAY, B. S. and B. W. Roberts 2008. The emergence of metalworking. In Jones A. (ed), *Prehistoric Europe. Theory and Practice*. London, Blackwell, pp. 193–225.

OTTAWAY, B. and S. Seibel 1998. Dust in the Wind. Experimental casting of bronze in sand moulds. In Frère-Sautot M.-C. (ed.), *Paléométallurgie des cuivres (Actes du colloque de Bourg-en-Bresse et Beaune, 17–18 oct. 1997)*. Montagnac, Monique Mergoil, pp. 59–63.

PANEI, L., G., Rinaldi and M. Tosi 2005. Investigations on ancient beads from the Sultanate of Oman (Ra's al-Hadd, Southern Oman). *ArcheoSciences (Revue d'Archéométrie)* 29, 151–155.

PEAKE, H. 1928. The copper mountain of Magan. *Antiquity* 2.8, 452–457.

PERNICKA, E. 1990. Gewinnung und Verbreitung der Metalle in prähistorischer Zeit. *Jahrbuch des Römisch-Germanischen Zentralmuseums, Mainz* 37.1, 21–129.

PERNICKA, E., F. Begemann, S. Schmitt-Strecker and A. P. Grimanis 1990. On the composition and provenance of metal artefacts from Poliochni on Lemnos. *Oxford Journal of Archaeology* 9.3, 263–298.

PERRY, W. J. 1937. *The Growth of Civilization* (2nd edition). Harmondsworth, Penguin Book.

PICKIN J. and S. Timberlake 1988. Stone hammers and firesetting. *Bulletin of the Peak District Mines Historical Society Ltd* 10.3, 165–167.

PIGOTT, V. C. 1999. The Development of metal production on the Iranian Plateau. In V. C. Pigott (ed.), *The Archaeometallurgy of the Asian Old World*. Philadelphia, University of Pennsylvania Museum, pp. 73–106.

POSSEHL, G. L. 1996. Meluhha. In Reade J. (ed.), *The Indian Ocean in Antiquity*. London / New York, Kegan Paul International, pp. 133–208.

PRANGE, M. 2001. 5000 Jahre Kupfer im Oman (Vergleichende Untersuchungen zur Charakterisierung des omanischen Kupfers mittels chemischer und isotopischer Analysenmethoden). *Metalla*, 8.2

PRANGE, M. K., H.-J Götze, A. Hauptmann and G. Weisgerber 1999. Is Oman the ancient Magan? Analytical Studies on Copper from Oman. In Young S. M. M., A. M. Pollard, P. Budd and R. A. Ixer (eds.), *Metals in Antiquity* (BAR International Series 792). Oxford, Archaeopress, pp. 187–192.

PRANGE, M. and A. Hauptmann 2001. The chemical composition of bronze objects from 'Ibrī/Selme. In Yule P. and G. Weisgerber (eds.), *The Metal Hoard from 'Ibrī/Selme Sultanate of Oman* (Prähistorische Bronzefunde 20.7). Stuttgart, Franz Stainer Verlag, pp. 75–84.

PULAK, C. 2000, The Copper and Tin Ingots from the Late Bronze Shipwreck at Uluburun. In Yalçin Ü. (ed.), *Anatolian Metal I* (Veröffentlichungen aus dem Deutschen Bergbau-Museum Bochum 92. / Anschnitt., Beiheft 13). Bochum, Bergbau-Museum, pp. 137–157.

QANDIL, H. 2005. Survey and excavations at Saruq Al-Hadeed, 2002–2003. In Hellyer P. and M. Ziolkowski (eds.), *Proceedings of the 1st Annual Symposium on recent Palaeontological & Archaeological discoveries in the Emirates, Al Ain 2003* (Emirates Heritage 1). Al Ain, Zayed Center for Heritage and History, pp. 121–139.

RAI, A. K. and S. S. Das 2011, Late Quaternary changes in surface productivity and oxygen minimum zone (OMZ) in the northwestern Arabian Sea, Micropaleontologic and sedimentary record at ODP site 728A. *Journal of Earth System Science* 120.1, 113–121.

RAPP, G. and S. Swiny 2003. Introduction. In Swiny S., G. Rapp and E. Herscher (eds.), *Sotira Kaminoudhia. An Early Bronze Age Site in Cyprus* (American Schools of Oriental Research archaeological reports 08 / Cyprus American Archaeological Research Institute monograph series 4). Boston MA, American Schools of Oriental Research, pp. 1–8.

REINHARDT, B. M. 1969. On the genesis and emplacement of ophiolites in the Oman Mountains geosyncline. *Schweizerische Mineralogische und Petrographische Mitteilungen* 49.1, 1–30.

ROGERS, W. 1895. *Outlines of the History of Early Babilonia*. Leipzig, Stauffer

ROTHENBERG, B. 1978. Excavations at Timna Site 39. A Chalcolithic Copper Smelting Site and Furnace and its Metallurgy. In Rothenberg B. and R F Tylecote (eds.), *Chalcolithic Copper Smelting* (Institute for Archaeo-Metallurgical Studies / Archaeometallurgy IAMS monograph). London, Institute for Archaeo-Metallurgical Studies, pp. 1–15.

RYNDINA, N. V. and L. K. Yakhontova 1985. The earliest copper artefact from Mesopotamia. *Sovetskaja Archeologija* 2, 155–165.

SCOTT, D. A. 1991. *Metallography and Microstructure of Ancient and Historic Metals*. Marina del Rey, CA, Getty Conservation Institute / Archetype Books.

SMITH, C. S. 1981. *A Search for Structure. Selected Essays on Science, Art, and History*. Cambridge MA, MIT Press.

STOS-GALE, Z. A., N. H. Gale, J. Houghton and R. Speakman 1995. Lead isotope analyses of ores from the Western Mediterranean. *Archaeometry* 37.2, 407–415.

THORNTON, C. P. 2007. Of brass and bronze in prehistoric Southwest Asia. In La Niece S., D. Hook, P. T. Craddock (eds.), *Metals and Mines. Studies in Archaeometallurgy*. London, Archetype Publications / The British Museum, pp. 189–201.

THORNTON, C. P. 2016. Excavations at Kasr al-Khafaji (Tower 1146) (Chapter 3). In Thornton C. P., C. M. Cable and G. L. Possehl (eds.), *The Bronze Age Towers at Bat, Sultanate of Oman. Research by the Bat Archaeological Project, 2007–12*. Philadelphia, University of Pennsylvania Press, pp. 25-47.

THORNTON, C. P., C. M. Cable and G. L. Possehl 2013. Three seasons at Kasr al-Khafaji (Tower 1146) at Bat, Oman. In Frenez D. and M. Tosi (eds.), *South Asian Archaeology 2007, Vol. 1. Prehistoric Periods* (BAR International Series 2454). Oxford, Archaeopress, pp. 255–268.

THORNTON, C. P. and C. B. Ehlers 2003. Early brass in the ancient Near East. *Institute for Archaeo-Metallurgical Studies Journal* 23, 3–8.

THORNTON, C. P. and R. O. Ghazal 2016. Typological and Chronological Consideration of the Ceramics at Bat, Oman (Chapter 9). In Thornton C. P., C. M. Cable and G. L. Possehl (eds.), *The Bronze Age Towers at Bat, Sultanate of Oman. Research by the Bat Archaeological Project, 2007-12*. Philadelphia, University of Pennsylvania Press, pp. 179-216.

THORNTON, C. P. and C. Giardino 2012. Serge Cleuziou and the 'Tin Problem'. In Giraud J. and G. Gernez (eds.), *Aux Marges de l'Archéologie. Hommage a Serge Cleuziou* (Travaux de la Maison René-Ginouvès 16). Paris, De Boccard, pp. 253–260.

TOSI, M. 1975. Notes on the Distribution and Exploitation of Natural Resources in Ancient Oman. *Journal of Oman Studies* 1, 187–206.

TOSI, M. 1976. The Dating of Umm an-Nar Culture and a Proposed Sequence for Oman in Third Millennium BC. *Journal of Oman Studies* 2, 81– 92.

TOSI, M. 1989. Protohistoric archaeology in Oman. The first thirty years (1956–1985). In Costa P. M. and M. Tosi (eds.), *Oman Studies. Papers on the Archaeology and History of Oman* (Serie Orientale Roma LXIII). Rome, Istituto Italiano per il Medio ed Estremo Oriente, pp. 135–161.

TOSI, M. 1998. La missione archeologica nel Sultanato di Oman. In *Missioni Archeologiche Italiane. La Ricerca Archeologica Antropologica Etnologica*. Rome, "L'Erma" di Bretschneider, pp. 231–234.

TOSI, M. and D. Usai. 2003, Preliminary reports on the excavations at Wadi Shab, Area 1, Sultanate of Oman. *Arabian Archaeology and Epigraphy* 14, 8–23.

USAI, D. 2005. Chisels or perforators? The lithic industry of Ras al-Hamra 5 (Muscat, Oman). *Proceedings of the Seminar for Arabian Studies* 35, 1–9.

USAI, D. 2006. A fourth–millennium Be Oman site and its context, Wadi Shab GAS-1. *Proceedings of the Seminar for Arabian Studies* 36, 275–288.

VOSMER, T. 2007. Early Bronze Age navigation and trade routes. In Cleuziou S. and M. Tosi (eds.), *In the Shadow of the Ancestors. The Prehistoric Foundations of the Early Arabian Civilization in Oman*. Muscat, Ministry of Heritage and Culture of the Sultanate of Oman, pp. 207–209.

VOSMER, T. 2008. Shipping in the Bronze Age, How large was a 60-gur Ship?. In Olijdam E. and R. H. Spoor (eds.), *Intercultural Relations between South and Southwest Asia. Studies in commemoration of E.C.L. During Caspers (1934–1996)* (BAR Int. Series 1826). Oxford, Archaeopress, pp. 230–235.

WALDBAUM, J. C. 1980. The First Archaeological Appearance of Iron and the Transition to the Iron Age. In Wertime T. A. and J. D. Muhly (eds.), *The Coming of the Age of Iron*. New Haven, Yale University Press, pp. 69–98.

WEEKS, L. R. 1997. Prehistoric metallurgy at Tell Abraq, United Arab Emirates. *Arabian Archaeology and Epigraphy* 8, 11–85.

WEEKS, L. R. 2000. Metal artefacts from the Sharma tomb. *Arabian Archaeology and Epigraphy* 11, 180–198.

WEEKS, L. R. 2003. *Early metallurgy of the Persian Gulf. Technology, trade, and the Bronze Age world.* Boston, Brill.

WEEKS, L. R. 2008. Coals to Newcastle, copper to Magan? Isotopic analyses and the Persian Gulf metal trade. In La Niece S., D. Hook, P. T. Craddock (eds.), *Metals and Mines. Studies in Archaeometallurgy.* London, Archetype Publications / The British Museum, pp. 89–96.

WEISGERBER, G. 1977. Beobachtungen zum alten Kupferbergbau im Sultanat Oman, Der Anschnitt. *Zeitschrift für Kunst und Kultur im Bergbau* 29.5-6, 190–211.

WEISGERBER, G. 1978. Evidence of Ancient Mining Sites in Oman. Preliminary Report. *Journal of Oman Studies* 4, 15–28.

YULE, P. A. and G. Weisgerber 2015. Al-Wasit Tomb W1 and other Sites. Redefining the Second Millennium BCE Chronology in South-Eastern Arabia. In Yule P. A. (ed.), *Archaeological Research in the Sultanate of Oman. Bronze and Iron Age Graveyards* (Expedition of the Deutsches Bergbau-Museum Bochum in Oman 1; Anschnitt Beiheft 28; Veröffentlichungen aus dem Deutschen Bergbau-Museum Bochum 208). Bochum, Bergbau-Museum, pp. 9–108.

Index

Aarja, 9, 10
AAS (Atomic Absorption Spectroscopy), 14, 17, 138
abrasive, 122
Abu Dhabi (Emirate), 27, 61, 63, 116
Acacia, 1, 6, 100, 120
Acacia ehrenbergiana, 6, 120
Acacia savannah, 1
Achaemenid Empire, 114
Ad-Dakhiliyah, 11
Afghanistan, 22, 83
Africa, 24, 25
Ag (silver), 13, 14, 33, 44, 45, 49, 61, 62, 101, 139-152. *See also* silver
Akkad, 20-25, 28
Al Ain, 61
Al-Ajal, 10
Al-Aqir, 100, 101, 102
Alashiya, 106
Al-Batinah, 3
Al-Beyda, 84
Alexander the Great, 114
Al-Hajar Mountains, 1, 7, 9, 22, 62, 98
Al-Lushal, 87
Al-Midra Ash-Shamali, 129
Al-Moyassar, 4, 22, 27, 41, 63, 76, 79, 84-88, 96-105, 108, 111, 116. *See also* Maysar
Al-Qusais, 108
Al-Ṣafāʾ, 116. *See also* ʿUqdat Al-Bakrah
Al-Sayab, 84
Al-Sufouh, 79
Aluminum (Al), 139
Al-Wasit, 108, 112, 113
Amarna, 106
Ambelikou-Aletri, 106
Amlah, 41
Anadara ehrenberghi, 83
Anatolia, 13, 23, 27, 38, 83, 84
anklets, 137
An-Nasiriyah, 22
annealing, 13-16, 38, 55, 58, 67, 69, 122, 133

anthropomorphic figures, 100
antimony (Sb), 13, 14, 143, 161. *See also* Sb (antimony)
anvils, 58, 66, 122
Aqabah, 95
argillite, 122
Arja, 27, 84, 87, 116
arrowheads, 114, 116, 137
arsenic (As), 10-13 28, 31, 33, 36, 37, 43, 44, 49, 52, 54, 62, 69, 72-80, 85, 93, 100, 101, 108, 110, 138, 141-144, 156, 157, 160. *See also* As (arsenic)
arsenical copper, 12, 13, 36, 44, 70, 110, 117
arsenopyrite, 31
As (arsenic), 13, 22, 30, 31, 33, 36, 38, 44, 45, 49, 52, 54, 62, 72-77, 80, 88, 96, 101, 108, 110, 117, 139-152. *See also* arsenic (As)
Assayab, 96
As-Suwayh, 30
Atomic Absorption Spectroscopy (AAS), 14. *See also* AAS (Atomic Absorption Spectroscopy)
awls, 3, 20, 30, 31, 33, 36, 37, 41-46, 49, 50, 57, 61, 62, 70-74, 79, 108, 110, 141-150
axes, 9, 38, 70, 74-76, 99, 100, 114, 116, 117, 121, 122, 125-127, 148, 149
azurite, 9, 11

Ba (barium), 139.
Badakhshan, 22
Bahla, 41, 100
Bahrain, 21-23, 27, 28, 100
Baluchistan, 23, 38, 83
Bandar Abbas, 25
bangles, 11, 108, 114, 136-138
bar, 34, 49, 50, 67, 68, 125
basalt, 7, 116
Basel, 87, 89, 91
Bat, 2, 41, 62-64, 80
Batin, 27, 96

Baushar, 108
Bayda, 9
beads, 30, 31, 41, 58, 77, 80, 110, 114
bellows, 94, 99
beneficiation (process), 91-93
Bid Bid, 11, 86, 88
bifid tool, 50
Bilad Al-Maidin, 10, 87-92, 96-98, 116
Bir Kalher, 111
bismuth (Bi), 36, 44, 93
Bisyah, 41
bitumen, 23, 66
black-painted red ware, 63
blades, 52, 74, 76, 77, 80, 122, 125, 131
blocklets, 43-45, 49, 54, 56, 77, 79, 122-125, 141-144, 147-150
bloom, 125
blowpipes, 94
Bochum, 27, 84, 99.
boomerang tool, 49
Borgo di Gela, 127
bornite, 9
bowls, 22, 114, 116, 122, 133, 134, 137
brass, 12, 13, 65
brochantite, 9, 101, 102
bronze, 11-13, 30, 31, 44, 63, 69, 70, 73, 76-80, 83, 99, 108, 113, 114, 117, 118, 122, 125, 129, 132, 138-141, 144
Bronze Age, 3, 5, 7, 12, 20, 22, 24, 27, 36, 40, 41, 62-66, 72, 80, 84, 88, 96-102, 106, 111, 112, 115, 130, 131, 141, 142
Buraimi, 41, 108

Ca (calcium), 65, 83, 139, 152. *See also* calcium (Ca)
cache, 137
calcite, 77, 80, 114
calcium (Ca), 83, 141-144. *See also* Ca (calcium)
Calligonum crinitum, 6

camel, 6, 93, 120
canes, 94
carbon (C), 12, 93, 120
carnelian, 21, 114, 136
casting, 13-15, 27, 44, 55, 63, 67-70, 77, 79, 84, 85, 100, 108, 110-113, 116-122, 125, 127, 137, 142, 149, 150
Central Asia, 83
chalcopyrite, 9, 31, 85, 99, 106, 115, 122
charcoal, 93, 95, 99, 100, 118, 120, 125, 127
chimney, 99, 120
chisels, 30-33, 36, 43-47, 50, 54, 57, 58, 66, 67, 73, 77, 79, 86, 88, 89, 110, 132, 134, 143-150
chlorine (Cl), 83, 141. *See also* Cl (chlorine)
chlorite, 22, 30, 31
chrysocolla, 9
cipraea, 3
Cl (chlorine), 83, 139, 152. *See also* chlorine (Cl)
cleaver, 100
Co (cobalt), 9, 10, 101, 139, 141, 142, 145-152. *See also* cobalt (Co)
coal, 93
cobalt, 10, 28, 101, 141-143, 155, 160
cold working, 44, 55, 57, 69
Conus sp., 3
copper sulfide, 83, 99, 101, 144
Cornulaca arabica, 6
cosmetics, 83, 144, 152
cows, 93
Cr (chromium), 139, 152
crochets, 48, 50, 73, 74, 148, 149
crucibles, 17, 22, 27, 38, 62, 63, 67-70, 77, 79, 96, 99, 100, 116, 120, 122, 139, 144
cuneiform texts, 20, 22, 23, 41
cutters, 31, 52
Cyperus conglomeratus, 6
Cyprus, 7, 23, 27, 102, 106, 107, 115

Daba, 114, 116, 136, 137. *See also* Dibba
daggers, 3, 43-45, 52, 53, 60, 61, 65, 77, 108, 112-114, 118, 137, 138, 142, 148
Dalbergia sp., 100
dams, 100
Dank, 116
dates, 11, 30, 41, 63, 66, 96, 106, 116, 144
dendrite (structure), 14, 55
De Re Metallica, 87, 89, 91

Deutsches Bergbau Museum (Bochum), 27, 84, 99, 111
Dhayah, 108
Dhofar, 1-3, 9, 11
Dibba, 116. *See also* Daba
Dilmun, 20-25, 28
Dimtu, 22
diorite, 21, 25
Dipterygium glaucum, 6
dolphins, 41
drills, 31, 33, 36, 37, 48, 50, 74, 143, 145, 146, 149
Dubai (Emirate), 108, 116
dugongs, 63
dung, 93, 120

earrings, 30, 76, 77
EDXRF (Energy Dispersive X-Ray Fluorescence), 139
Egypt, 13, 24, 25, 29, 58, 94, 106, 114
electrum, 12, 13
Empty Quarter, 1, 5
Energy Dispersive X-Ray Fluorescence (EDXRD), 139. *See also* EDXRF (Energy Dispersive X-Ray Fluorescence)
Enkomi, 106
enstatite, 58
Ethiopia, 24
ethnology, 125
Euphrates (River), 20, 21
Europe, 38, 43, 87-91, 96, 100, 125

Fahal Island, 33
fahlertz, 36
farmlands, 41
Fe (iron), 9, 13, 14, 44, 45, 55, 65, 83, 101, 110, 138-152. *See also* iron
feldspar, 5
fireplaces, 67, 99
fire setting technique, 87
fishhooks, 13, 30, 31, 36, 46-49, 61, 65, 66, 70-72, 108, 143, 144. *See also* hooks
fishing, 3, 31, 33, 34, 36, 41, 43, 47, 49, 50, 54, 57, 58, 61, 63, 66, 74, 84, 108
flat axes, 74-76, 99, 100, 148, 149
flint, 30, 31, 41, 84, 91, 98
flux, 91, 93, 99, 104
forges, 120, 122, 125
forging, 44, 125
forks, 43
fuel, 1, 91, 93, 95, 100, 118
furnaces, 1, 13, 58, 92-99, 104, 105, 111, 116, 118, 120, 122, 125, 129, 131

gabbro, 7, 9, 25
galena, 10
Galilee, 58
Ganeshwar, 38
gangue, 87, 91-93, 104
garlic, 85
GAS-1 (Wadi Shab), 30-32, 38, 141-145, 151-162
Gebel Saleli, 88
Georgius Agricola, 87, 89, 91
Ghalilah, 108
Ghanadha, 80
Ghar Chale, 84
Girsu, 22, 23
goethite, 9, 125
gold (Au), 10-14, 20, 92, 116, 136
gossan, 9, 85
granulation (technique), 137
Gudea of Lagash, 20, 25, 26
Gujarat, 23
Gulf of Oman, 1, 2, 7

Hafit type graves, 40, 41, 61
hafting, 86, 89, 127, 129, 131
Hala Sultan Tekke, 106
halberds, 114
hammering, 3, 14, 16, 30, 31, 33, 37, 38, 43, 44, 49, 54, 56, 57, 63, 74, 77, 100, 108, 118, 125, 127
hammers, 58, 85-89, 96, 99, 116, 122, 125, 127-129
hammer-stones, 36, 66, 86, 89, 92
Hammurabi, 23
Harappan (Civilization), 23, 38, 58, 66, 73, 150. *See also* Indus Civilization, Meluhha
Harvard Archaeological Survey, 27, 84
Hawasina Napples, 7, 10
Hawqayn, 116
HD-1 (Ras Al-Hadd), 101-103
HD-6 (Ras Al-Hadd), 3, 36, 41-61, 69-74, 84, 101, 141-144, 147, 151-162
HD-7 (Ras Al-Hadd), 79, 81
He (helium), 25, 27, 84, 139. *See also* helium (He)
helium (He), 139. *See also* He (helium)
hematite, 10, 11, 93, 104, 125
Hili, 23, 61, 63, 80
hoes, 77, 86, 100, 129-131
Holocene, 2, 29, 66
hooks, 3, 31, 37, 43, 47, 49, 54, 58, 65, 72, 141, 142, 145-150. *See also* fishhooks
Hormuz, 1, 21, 25

hot working, 44, 57
Huqain, 88
huts, 30

Ibra, 27, 63, 96
Ibri, 41, 108, 129
Ibri/Selme (hoard), 136, 137. *See also* Selme hoard
India, 23, 27, 28, 66, 72, 100
Indian Ocean, 23, 66
Indus Civilization, 23, 72. *See also* Harappan Civilization, Meluhha
Indus script, 66
Indus Valley, 12, 22, 23, 27, 29, 38, 58, 63, 66, 106
ingots, 22, 27, 43, 50, 54, 57, 67, 68, 69, 77-80, 99-103, 106, 117, 122-125, 148, 149
Iran, 20, 22, 25, 27, 29, 38, 83, 84, 99, 106
Iraq, 20, 66
iron (Fe), 9-14, 31, 36, 55, 65, 69, 72, 73, 83, 86, 88, 93, 98-110, 114-117, 120-125, 131-133, 136-142, 149, 155. *See also* Fe (iron)
Iron Age, xvii, 6, 11, 65, 88, 98, 99, 108, 114-117, 125, 131, 133, 136, 137
iron oxides, 9, 93, 122
iron sulfide, 9
irrigation, 29, 41
Isin-Larsa, 20
Islamic period, 88-98
Israel, 84

Ja'alan, 30, 40, 41, 63, 108
Jabal Al-Hawrah, 114
Jabal Al-Ma'adan, 27
jarosite, 9
Jebel Al-Akhdar, 3
Jebel Al-Emalah, 58
Jebel Saleli, 116
Jebel Salim Khamis, 66
Jemdet Nasr, 41
jewelry, 31, 80, 116, 136

K (potassium), 139, 152. *See also* potassium (K)
Kalopsidha Koufos, 106
Kasr Al-Khafaji, 62. *See also* Tower 1146
Khashbah, 98
Khudra, 108
knives, 33, 34, 36, 50, 74, 75, 148-150
kohl, 83
Kozlu, 84

Lagash, 20, 22, 25, 26
lagoons, 2, 30, 41
Larsa, 20, 21
Lasail, 9, 10, 27, 84, 87
lead (Pb), 11-14, 17, 28, 52, 69, 77, 102, 110, 141-144, 161. *See also* Pb (lead)
Lead Isotope Analysis, 14, 17, 102
leather, 49, 94
lime, 83, 144
Limeum arabicum, 6
limonite, 9, 93, 105, 125
limpets, 34
linen, 87
Lizq, 114, 136, 137
Luristan, 114
Luzaq, 27

Maerua, 100
Magan, xvii, 20-28, 41, 106
magnetite, 93, 125
malachite, 9, 11
manganese (Mn), 10, 141. *See also* Mn (manganese)
mangroves, 2
Masirah (island), 7, 9, 11, 55, 106, 108, 129, 142
Masirah Ophiolite, 9, 11
Masjad, 27
Matariya, 62. *See also* Tower 1147
matte, 99, 102
Maysar, 4, 22, 85-87, 98. *See also* Al-Moyassar
Medinat Hamad, 22
Mediterranean Sea, 7, 100, 102, 106, 113, 114
Mehrgarh, 38
melting, 3, 12-14, 38, 43, 44, 62, 63, 93, 108, 113, 118, 120, 125
Meluhha, 20, 21, 23, 28. *See also* Indus Civilization, Harappan Civilization
Mesopotamia, xvii, 13, 20-29, 38, 41, 61, 66, 106, 136
meteoric iron, 13, 65
Middle East, 13, 29
millstones, 58
mines, 3, 5, 22, 27, 84-88, 92, 101, 107, 115. *See also* mining
Minimum Detection Limit (MDL), 139, 141, 145-152
Minimum Relative Error (MRE), 139, 145-152
mining, 9, 10, 21, 27, 38, 84-89, 92, 106, 115, 137. *See also* mines
Mn (manganese), 139. *See also* manganese (Mn)

Mohenjo-Daro, 23
molds, 22, 69, 79, 99
mollusks, 2, 36, 50
monsoon, 1, 2, 41
mortars, 87, 92, 96, 97, 116
MRE (Minimum Relative Error), xiii, 139-152
mud-bricks, 41, 66, 99
Mullaq, 87, 96
multifunctional tool, xvii, 3, 34, 36, 38, 50
Musandam, 3, 106, 114
Muscat, 2, 10, 30, 38, 89
Musfa, 116
mussels, 34

Naram-Sin of Akkad, 25
Nasariyah, 100
native copper, 20, 38, 39, 141
Near East, xvii, 12, 22, 84, 87, 95, 96, 100, 104, 106, 113, 114, 136
needles, 3, 41, 48, 50, 58, 70, 73, 74, 99, 108, 110, 142, 150
Neem-Ka-Thana, 38
Neolithic period, 12, 20, 31, 38, 47, 55, 84
net-sinkers, 30
netting tools, 31, 33, 48, 50, 58, 70, 73, 74, 145, 146
Ni (nickel), 9, 10, 13, 31, 33, 36, 44, 45, 49, 50, 52, 54, 61, 62, 72, 73, 77, 83, 101, 108, 110, 139, 141-152. *See also* nickel (Ni)
nickel (Ni), 10, 13, 27, 28, 31, 33, 36, 43, 44, 49-55, 61, 62, 65, 69, 72-74, 77, 79, 83, 100, 101, 108, 110, 112, 117, 138, 141-144, 158, 159, 160. *See also* Ni (nickel)
Nizwa, 41, 88, 114, 115, 131, 136
Nubia, 24
Nujum, 86, 88, 89, 92

oases, 1, 2, 41, 43, 58, 62, 116, 117
obsidian, 20
Old Babylonian period, 20
olivine-gabbro, 25
open-cast mining, 84
open-pit mines, 87
optical microscope, 14
ores, 5, 7, 9-14, 17, 20, 21, 25-31, 36, 38, 44, 52, 55, 62, 63, 80, 83-106, 108, 111, 112, 116, 117, 122, 125, 138, 142
ornaments, 30, 50, 66, 80, 136
ostrea, 3
oysters, 34
oyster-shucking knife, 36, 50

177

Pakistan, 100
Palladium (Pd), 139. *See also* Pd (palladium)
patina, 36, 141, 142
Pb (lead), 13, 14, 44, 45, 77, 139-152. *See also* lead (Pb)
Pd (palladium), 139. *See also* Palladium (Pd)
pendants, 66
perforators, 30, 31, 33
peridotite, 7
pestles, 87, 89, 92
picks, 86, 87, 89
pigments, 83
pillow lavas, 7, 9, 106
pins, 33, 36, 37, 43-45, 49-52, 56, 61, 62, 76, 79, 141, 142, 145-150
Pistacia, 100
plano-convex ingots, 22, 27, 99, 100-102, 124
pointed tools, 33, 36, 37, 49, 50, 74, 89, 143-147
polishing (paste), 122
Politiko Phorades, 106
potassium (K), 139, 141. *See also* K (potassium)
prills, 62, 63, 96, 99, 104, 122
Prosopis cineraria, 6, 100, 116, 120
punches, 47, 50, 141
Pyrgos-Mavrorachi, 106
pyrite, 9, 106
pyritic copper ore, 99
pyrolusite, 83
pyrotechnology, 9, 58

Qal'at Al-Bahrain, 22, 100
Qarn Al-Muallaq, 116
Qarn Bint Saud, 125
Qattarah, 108
quartz, 5, 11, 25, 93, 122
Quasys, 125
Quriyat, 41, 89
Qurum, 38

Rajasthan, 38
Raki, 9, 87, 115, 116
rapiers, 108
Ras Al-Hadd, 3, 4, 9, 36, 41-61, 65, 66, 70, 72, 79-84, 100, 101, 103, 141-144, 147, 148, 150-162
Ras Al-Hamra, xvii, 3, 30, 33-38, 50, 141, 146, 151-162
Ras Al-Jinz, 3, 13, 23, 30, 41, 61-84, 108-112, 141-144, 148, 150-163
Rawdah, 111
razors, 76-80, 108, 114, 131, 132, 141, 144, 149, 150

Rb (rubidium), 139
recycling, xvii, 54, 55, 73-77, 80, 100, 108, 110, 117, 118, 122, 144
Red Sea, 95
reducing atmosphere, 55, 93, 105
refractory clay, 69, 94, 122
RH-5 (Ras Al-Hamra), 38, 50
RH-10 (Ras Al-Hamra), xvii, 30-38, 41, 50, 141-146, 151-162
rings, 41, 66, 76-80, 100, 108, 114, 136, 144, 149, 150
rivets, 41, 52, 61
RJ-1 (Ras Al-Jinz), 30, 108-112, 141-144, 150-161
RJ-2 (Ras Al-Jinz), 13, 63-84, 108, 141-163
RJ-6 (Ras Al-Jinz), 61, 66, 141, 142, 150, 153-162
RJ-10 (Ras Al-Jinz), 41, 66
RJ-13 (Ras Al-Jinz), 66
RJ-20 (Ras Al-Jinz), 66
roasting, 93, 122
robbing, 137
rods, 49
ropes, 34
Rub Al-Khali, xvii, 1, 5, 6, 116-118, 120, 122
Rumaylah, 125, 129
Rustaq, 116

S (sulfur), 43, 70, 83, 101, 139, 144, 152. *See also* sulfur (S)
Saar, 22, 100
Sachrut Al-Hadri, 108, 129
Samad, 5, 27, 58, 84, 98, 108, 111
Samdah, 84
sand, 1, 5, 6, 41, 69, 116, 117, 118, 120, 122, 123, 129
sandstone, 57
sardines, 33
Sardinia, 102
Sargon of Akkad, 20-25
Saruq Al-Hadid, 116, 125
Saudi Arabia, 21, 31, 102, 129
Sb (antimony), 13, 14, 77, 139-151. *See also* antimony (Sb)
Scanning Electron Microscope, 14, 43, 80, 101. *See also* SEM
scrapers, 52
scraps, 54, 122
Se (selenium), 139, 152
seals, 65, 66, 114, 125, 150
seashells, 31, 34, 47. *See also* shells
Selme (hoard), 11, 125, 136-138
SEM, 14, 17, 18, 56, 77-82, 101, 104. *See also* Scanning Electron Microscope

Semail Ophiolite, 7-10, 28, 84, 98
Semdah, 9, 87
semi-finished (objects), 49, 52, 54, 56, 67, 77, 110, 120, 121, 125, 141, 157, 159, 163
shaft-hole hoes, 100, 125, 127, 129
shafts, 87
Sharm, 142
shell, 3, 30, 36, 43, 58, 63, 70, 83, 143, 148-152. *See also* seashells
shellfish, 34
shell-opener tools, 36, 50, 58, 70, 74, 143
Shimal, 108, 142
shovels, 86
Si (silicon), 139
Sicily, 127
siderite, 125
silver (Ag), 11-14, 20, 31, 33, 36, 37, 49, 61, 62, 108, 116, 117, 136, 141-143, 154. *See also* Ag (silver)
Sinai Peninsula, 27
sinkers, 30
sinking, 14
slags, 10, 27, 62, 63, 84, 88, 92-107, 111, 112, 115, 116, 122
smelting, 12, 13, 22, 27, 38, 43, 44, 55, 63, 83, 84, 88, 91-107, 111, 115, 116, 118, 122, 125, 137
Sn (tin), 13, 14, 31, 44, 45, 73, 76, 113, 114, 138, 139, 141, 144-151
soapstone, 30, 41, 49, 58. *See also* tin (Sn)
softstone vessels, 114
Sohar, 9, 27, 106
spearheads, 108, 112, 113
sprues, 122
Sr (strontium), 139, 152. *See also* strontium (Sr)
stakes, 129, 130
stannite, 31
steatite, 22, 30, 31, 58
steel, 12, 13
stone-hammers, 87, 88, 89
stone tools, 122
Strait of Hormuz, 1, 21
strombus, 3
strontium (Sr), 141. *See also* Sr (strontium)
Sudan, 24
sulfides, 7, 9, 10, 83, 85, 93, 99, 101, 106, 144
sulfur (S), 83, 93, 99, 100, 101, 122. *See also* S (sulfur)
Sumer(ian), xvii, 5, 20, 21-25, 27, 41
Sur, 3, 30, 41
surface mining, 87

Susa, 25
Swiss-knife, 33, 36
Switzerland, 87
swords, 108, 112, 113
SWY-2 (As-Suwayh), 30
Syria, 27, 29, 106

Tajikistan, 83
Tal-i-Iblis, 38
Tamarix, 100
tang, 34-36, 50, 131
tapping, 95, 96, 99, 107
Tawi Raki, 87, 115
Tawi Ubaylah, 88, 96
Tell Abraq, 76, 79, 80, 173
Tell Magzaliya, 20
Tepe Ghabristan, 38
Thermoluminescence (TL), 96
Third Dynasty of Ur, 23
Ti (titanium), 139. See also titanium (Ti)
Tigris River, 20, 21
Timna, 84, 95
Tin (Sn), 12-14 20, 31, 33, 44, 63, 69, 70, 73-80, 83, 106, 108, 113, 114, 117, 125, 138, 141-144, 162, 163. See also Sn (tin)
titanium (Ti), 141
Tiwi, 30
Tower 1146 (Bat), 62. See also Kasr Al-Khafaji
Tower 1147 (Bat), 62. See also Matariya
Tower 1156 (Bat), 62, 63
Troodos Ophiolite, 7, 106
Turkey, 84, 106
turquoise, 20
turtles, 41, 63, 65
tuyères, 84, 94, 95

Uluburun, 106
Umm an-Nar (period), 3, 13, 22, 27, 40, 44, 58, 62-65, 74, 76, 79, 80, 84, 88, 96, 98, 100, 104, 106, 113, 137, 141, 143, 144
Umm an-Nar (island), 58, 65, 79
Umm an-Nar (graves), 40, 63, 137
Unar 2, 79
underground mining, 84, 87, 88
unfinished (objects), 30, 31, 33, 43, 47, 73, 122
United Arab Emirates, 27, 29, 63, 100
upwelling, 41
'Uqdat Al-Bakrah, 5, 6, 116-137
Ur, xvii, 20-23, 25
Ur-Enki, 22

Ur-Nanshe, 20
Uruk, 20, 28
Uzbekistan, 83

ventilation, 87, 91
Veshnoveh, 84
vessels, 22, 23, 38, 41, 77, 108, 114, 116, 117, 125, 127, 129, 133, 135-138

Wadi Aday, 33
Wadi Ahin, 27
Wadi Andam, 27
Wadi Bani Kharus, 11
Wadi Fulayj, 41
Wadi Ibra, 27
Wadi Jizzi, 27, 112
Wadi Miadin, 88, 116
Wadi Mijlas, 41
Wadi Musaw, 9
Wadi Nam, 96
Wadi Qatof, 116
Wadi Raki, 116
Wadi Sahl, 96, 111
Wadi Samad, 27, 98, 108, 111
Wadi Samah, 22, 99
Wadi Samail, 11, 88
Wadi Shab, 30-32, 38, 141, 145, 151-162
Wadi Shatahl, 9, 11
Wadi Sunaysl, 108
Wadi Suq (period), 22, 44, 63, 96, 106, 108, 110-114, 141, 142
weapons, xvii, xvii, 3, 20, 43, 47, 52, 57, 61, 108, 112-114, 117, 122, 125, 136, 138, 141, 157, 159, 163
whetstones, 57, 122
wood, 12, 20, 49, 50, 61, 86, 87, 93, 100, 118, 120, 125
workshop, xvii, 22, 27, 47, 63, 65, 66, 67, 80, 116-118, 122, 125, 127, 131, 137

X-Ray Diffraction (XRD), 14, 58. See also XRD (X-Ray Diffraction)
X-Ray Fluorescence (XRF), 14, 17, 19, 36, 43, 47, 83, 117, 139. See also XRF (X-Ray Fluorescence)
XRD (X-Ray Diffraction), 14, 17, 58. See also X-Ray Diffraction (XRD)
XRF (X-Ray Fluorescence), 14, 17, 19, 31, 43-45, 65, 66, 69, 70, 79, 112, 125, 139, 140, 142, 144, 151-163. See also X-Ray Fluorescence (XRF)

Y (yttrium), 139
Yanqul, 115
Yemen, 1

zinc (Zn), 11-14, 65, 69, 141, 142, 154. See also Zn (zinc)
Ziziphus, 100
Zn (zinc), 13, 14, 44, 45, 65, 139-152. See also zinc (Zn)
Zr (zirconium), 139
Zygophyllum mandavillei, 6